Review of the Department of Defense Research Program on Low-Level Exposures to Chemical Warfare Agents

Committee on Toxicologic Assessment of Low-Level Exposures to Chemical Warfare Agents

Committee on Toxicology

Board on Environmental Studies and Toxicology

Division on Earth and Life Studies

NATIONAL RESEARCH COUNCIL
OF THE NATIONAL ACADEMIES

THE NATIONAL ACADEMIES PRESS
Washington, D.C.
www.nap.edu

THE NATIONAL ACADEMIES PRESS 500 Fifth Street, NW Washington, DC 20001

NOTICE: The project that is the subject of this report was approved by the Governing Board of the National Research Council, whose members are drawn from the councils of the National Academy of Sciences, the National Academy of Engineering, and the Institute of Medicine. The members of the committee responsible for the report were chosen for their special competences and with regard for appropriate balance.

This project was supported by Contract No. W81K04-05-C-7006 and DAMD17-99-C-9049 between the National Academy of Sciences and the U.S. Army. Any opinions, findings, conclusions, or recommendations expressed in this publication are those of the author(s) and do not necessarily reflect the view of the organizations or agencies that provided support for this project.

International Standard Book Number 0-309-10021-6

Additional copies of this report are available from:

The National Academies Press
500 Fifth Street, NW
Box 285
Washington, DC 20055

800-624-6242
202-334-3313 (in the Washington metropolitan area)
http://www.nap.edu

Printed in the United States of America

THE NATIONAL ACADEMIES

Advisers to the Nation on Science, Engineering, and Medicine

The **National Academy of Sciences** is a private, nonprofit, self-perpetuating society of distinguished scholars engaged in scientific and engineering research, dedicated to the furtherance of science and technology and to their use for the general welfare. Upon the authority of the charter granted to it by the Congress in 1863, the Academy has a mandate that requires it to advise the federal government on scientific and technical matters. Dr. Ralph J. Cicerone is president of the National Academy of Sciences.

The **National Academy of Engineering** was established in 1964, under the charter of the National Academy of Sciences, as a parallel organization of outstanding engineers. It is autonomous in its administration and in the selection of its members, sharing with the National Academy of Sciences the responsibility for advising the federal government. The National Academy of Engineering also sponsors engineering programs aimed at meeting national needs, encourages education and research, and recognizes the superior achievements of engineers. Dr. Wm. A. Wulf is president of the National Academy of Engineering.

The **Institute of Medicine** was established in 1970 by the National Academy of Sciences to secure the services of eminent members of appropriate professions in the examination of policy matters pertaining to the health of the public. The Institute acts under the responsibility given to the National Academy of Sciences by its congressional charter to be an adviser to the federal government and, upon its own initiative, to identify issues of medical care, research, and education. Dr. Harvey V. Fineberg is president of the Institute of Medicine.

The **National Research Council** was organized by the National Academy of Sciences in 1916 to associate the broad community of science and technology with the Academy's purposes of furthering knowledge and advising the federal government. Functioning in accordance with general policies determined by the Academy, the Council has become the principal operating agency of both the National Academy of Sciences and the National Academy of Engineering in providing services to the government, the public, and the scientific and engineering communities. The Council is administered jointly by both Academies and the Institute of Medicine. Dr. Ralph J. Cicerone and Dr. Wm. A. Wulf are chair and vice chair, respectively, of the National Research Council.

www.national-academies.org

COMMITTEE ON TOXICOLOGIC ASSESSMENT OF LOW-LEVEL EXPOSURES TO CHEMICAL WARFARE AGENTS

Members

GARY P. CARLSON *(Chair),* Purdue University, West Lafayette, IN
A. JOHN BAILER, Miami University, Oxford, OH
JONATHAN BORAK, Yale School of Medicine, New Haven, CT
JANICE E. CHAMBERS, Mississippi State University, Mississippi State
MARION EHRICH, Virginia Tech, Blacksburg
LYNNE HABER, Toxicology Excellence for Risk Assessment, Cincinnati, OH
SANFORD S. LEFFINGWELL, HLM Consultants, Auburn, GA
DAVID H. MOORE, Battelle Eastern Science and Technology Center, Aberdeen, MD
CAREY POPE, Oklahoma State University, Stillwater
PETER S. SPENCER, Oregon Health & Science University, Portland
NGA LIEN TRAN, Exponent, Inc., Washington, DC

NRC Staff

KULBIR S. BAKSHI, Project Director
CAY BUTLER, Editor
AIDA C. NEEL, Program Associate
ALEXANDRA STUPPLE, Senior Editorial Assistant

COMMITTEE ON TOXICOLOGY

Members

BAILUS WALKER, JR. *(Chair)*, Howard University Medical Center and American Public Health Association, Washington, DC
MELVIN E. ANDERSEN, CIIT-Centers for Health Research, Research Triangle Park, NC
EDWARD C. BISHOP, Parsons Corporation, Fairfax, VA
GARY P. CARLSON, Purdue University, West Lafayette, IN
JANICE E. CHAMBERS, Mississippi State University, Mississippi State
LEONARD CHIAZZE, JR., Georgetown University, Washington, DC
JUDITH A. GRAHAM, American Chemistry Council, Arlington, VA
SIDNEY GREEN, Howard University, Washington, DC
MERYL KAROL, University of Pittsburgh, Pittsburgh, PA
STEPHEN U. LESTER, Center for Health Environment and Justice, Falls Church, VA
DAVID H. MOORE, Battelle Memorial Institute, Bel Air, MD
CALVIN C. WILLHITE, State of California, Berkeley, CA
GERALD WOGAN, Massachusetts Institute of Technology, Cambridge

Staff

KULBIR S. BAKSHI, Project Director
SUSAN N.J. MARTEL, Senior Program Officer
ELLEN K. MANTUS, Senior Program Officer
AIDA NEEL, Program Associate
TAMARA DAWSON, Senior Program Assistant
ALEXANDRA STUPPLE, Senior Editorial Assistant
SAM BARDLEY, Librarian

BOARD ON ENVIRONMENTAL STUDIES AND TOXICOLOGY[1]

Members

JONATHAN M. SAMET *(Chair)*, The Johns Hopkins University, Baltimore, MD
RAMÓN ALVAREZ, Environmental Defense, Austin, TX
THOMAS BURKE, The Johns Hopkins University, Baltimore, MD
JUDITH C. CHOW, Desert Research Institute, Reno, NV
RORY B. CONOLLY, CIIT-Centers for Health Research, Research Triangle Park, NC
COSTEL D. DENSON, University of Delaware, Newark
E. DONALD ELLIOTT, Willkie Farr & Gallagher LLP, Washington, DC
CHRISTOPHER B. FIELD, Carnegie Institution of Washington, Stanford, CA
SHERRI W. GOODMAN, Center for Naval Analyses, Alexandria, VA
JUDITH A. GRAHAM, American Chemistry Council, Arlington, VA
DANIEL S. GREENBAUM, Health Effects Institute, Cambridge, MA
ROBERT HUGGETT, Michigan State University, East Lansing
BARRY L. JOHNSON, Emory University, Atlanta, GA
JAMES H. JOHNSON, Howard University, Washington, DC
JUDITH L. MEYER, University of Georgia, Athens
PATRICK Y. O'BRIEN, ChevronTexaco Energy Technology Company, Richmond, CA
DOROTHY E. PATTON, International Life Sciences Institute, Washington, DC
STEWARD T.A. PICKETT, Institute of Ecosystem Studies, Millbrook, NY
JOSEPH V. RODRICKS, ENVIRON Corp., Arlington, VA
ARMISTEAD G. RUSSELL, Georgia Institute of Technology, Atlanta
MITCHELL J. SMALL, Carnegie Mellon University, Pittsburgh, PA
LISA SPEER, Natural Resources Defense Council, New York, NY
KIMBERLY M. THOMPSON, Harvard School of Public Health, Boston, MA
G. DAVID TILMAN, University of Minnesota, St. Paul
CHRIS G. WHIPPLE, ENVIRON Corp., Emeryville, CA
LAUREN A. ZEISE, California Environmental Protection Agency, Oakland

Senior Staff

JAMES J. REISA, Director
DAVID J. POLICANSKY, Scholar
RAYMOND A. WASSEL, Senior Program Officer for Environmental Sciences and Engineering
KULBIR BAKSHI, Senior Program Officer for Toxicology
EILEEN N. ABT, Senior Program Officer for Risk Analysis
K. JOHN HOLMES, Senior Program Officer
SUSAN N.J. MARTEL, Senior Program Officer
SUZANNE VAN DRUNICK, Senior Program Officer
ELLEN K. MANTUS, Senior Program Officer
RUTH E. CROSSGROVE, Senior Editor

[1]This study was planned, overseen, and supported by the Board on Environmental Studies and Toxicology.

OTHER REPORTS OF THE
BOARD ON ENVIRONMENTAL STUDIES AND TOXICOLOGY

Health Implications of Perchlorate Ingestion (2005)
Air Quality Management in the United States (2004)
Endangered and Threatened Species of the Platte River (2004)
Atlantic Salmon in Maine (2004)
Endangered and Threatened Fishes in the Klamath River Basin (2004)
Cumulative Environmental Effects of Alaska North Slope Oil and Gas Development (2003)
Estimating the Public Health Benefits of Proposed Air Pollution Regulations (2002)
Biosolids Applied to Land: Advancing Standards and Practices (2002)
The Airliner Cabin Environment and Health of Passengers and Crew (2002)
Arsenic in Drinking Water: 2001 Update (2001)
Evaluating Vehicle Emissions Inspection and Maintenance Programs (2001)
Compensating for Wetland Losses Under the Clean Water Act (2001)
A Risk-Management Strategy for PCB-Contaminated Sediments (2001)
Acute Exposure Guideline Levels for Selected Airborne Chemicals (4 volumes, 2000-2004)
Toxicological Effects of Methylmercury (2000)
Strengthening Science at the U.S. Environmental Protection Agency (2000)
Scientific Frontiers in Developmental Toxicology and Risk Assessment (2000)
Ecological Indicators for the Nation (2000)
Waste Incineration and Public Health (1999)
Hormonally Active Agents in the Environment (1999)
Research Priorities for Airborne Particulate Matter (4 volumes, 1998-2004)
The National Research Council's Committee on Toxicology: The First 50 Years (1997)
Carcinogens and Anticarcinogens in the Human Diet (1996)
Upstream: Salmon and Society in the Pacific Northwest (1996)
Science and the Endangered Species Act (1995)
Wetlands: Characteristics and Boundaries (1995)
Biologic Markers (5 volumes, 1989-1995)
Review of EPA's Environmental Monitoring and Assessment Program (3 volumes, 1994-1995)
Science and Judgment in Risk Assessment (1994)
Pesticides in the Diets of Infants and Children (1993)
Dolphins and the Tuna Industry (1992)
Science and the National Parks (1992)
Human Exposure Assessment for Airborne Pollutants (1991)
Rethinking the Ozone Problem in Urban and Regional Air Pollution (1991)
Decline of the Sea Turtles (1990)

Copies of these reports may be ordered from the National Academies Press
(800) 624-6242 or (202) 334-3313
www.nap.edu

OTHER REPORTS OF THE COMMITTEE ON TOXICOLOGY

Spacecraft Water Exposure Guidelines for Selected Contaminants, Volume 1 (2004)
Toxicologic Assessment of Jet-Propulsion Fuel 8 (2003)
Review of Submarine Escape Action Levels for Selected Chemicals (2002)
Standing Operating Procedures for Developing Acute Exposure Guideline Levels for Hazardous Chemicals (2001)
Evaluating Chemical and Other Agent Exposures for Reproductive and Developmental Toxicity (2001)
Acute Exposure Guideline Levels for Selected Airborne Contaminants, Volume 1 (2000), Volume 2 (2002), Volume 3 (2003), Volume 4 (2004)
Review of the US Navy's Human Health Risk Assessment of the Naval Air Facility at Atsugi, Japan (2000)
Methods for Developing Spacecraft Water Exposure Guidelines (2000)
Review of the U.S. Navy Environmental Health Center's Health-Hazard Assessment Process (2000)
Review of the U.S. Navy's Exposure Standard for Manufactured Vitreous Fibers (2000)
Re-Evaluation of Drinking-Water Guidelines for Diisopropyl Methylphosphonate (2000)
Submarine Exposure Guidance Levels for Selected Hydrofluorocarbons: HFC-236fa, HFC-23, and HFC-404a (2000)
Review of the U.S. Army's Health Risk Assessments for Oral Exposure to Six Chemical-Warfare Agents (1999)
Toxicity of Military Smokes and Obscurants, Volume 1(1997), Volume 2 (1999), Volume 3 (1999)
Assessment of Exposure-Response Functions for Rocket-Emission Toxicants (1998)
Toxicity of Alternatives to Chlorofluorocarbons: HFC-134a and HCFC-123 (1996)
Permissible Exposure Levels for Selected Military Fuel Vapors (1996)
Spacecraft Maximum Allowable Concentrations for Selected Airborne Contaminants, Volume 1 (1994), Volume 2 (1996), Volume 3 (1996), Volume 4 (2000)

Preface

In 1998, Congress directed the Secretary of Defense to review and modify U.S. Department of Defense (DOD) policies and doctrines that relate to protecting personnel from low-level exposure to chemical warfare agents (CWAs). In response to that congressional mandate, the Secretary of Defense directed DOD's Nuclear, Chemical, and Biological Defense Program to develop a research plan to obtain toxicologic and other data to assess health risks to U.S. forces. The data obtained from the proposed research would provide information to the Secretary of Defense to reassess policies and doctrines related to low-level exposures to CWAs. The research is intended to accomplish two objectives. The first is to support operational commanders in the field with information for real-time decision making required to accomplish their missions while not unduly jeopardizing the health and performance capability of their forces. The second is to understand, prevent, or reduce operationally relevant performance decrements, as well as the potential health consequences of low-level exposures that might not manifest immediately but could become evident months or years after exposure.

In response, DOD developed the multiyear research program on low-level exposures to chemical warfare agents entitled *Department of Defense Low-Level Chemical Warfare Agents (CWAs) Exposure Research Master Plan*. DOD requested that the National Research Council (NRC) review that research plan and comment on its adequacy and appropriateness, provide guidance on appropriate risk assessment methods for assessing toxicologic risk from low-level exposures to CWAs, and identify gaps and make recommendations for further research. The NRC convened the Committee on Toxicologic Assessment of Low-Level Exposures to Chemical Warfare Agents. The committee's report is intended to be helpful in focusing research efforts to improve operational management.

The DOD Low-Level CWA Exposure Research Master Plan (Research Plan) details the military's priorities for research needs and methods to address the effects of low-level agent exposure on operationally relevant performance in military personnel at the time of exposure and on potential delayed adverse health effects at some point after exposure. As stated in the Research Plan, both of these aspects of low-dose exposure "represent different points along the dose-response continuum—not separate problems."

This report has been reviewed in draft form by individuals chosen for their diverse perspectives and technical expertise, in accordance with procedures approved by the NRC Report Review Committee. The purpose of this independent review is to provide candid and critical comments that will assist the institution in making its published report as sound as possible and to ensure that the report meets institutional standards for objectivity, evidence, and responsiveness to the study charge. The review comments and draft manuscript remain confidential to protect the integrity of the deliberative process. We thank the following individuals for their review of this report: Barbara Callahan (University Research Engineers and Associates), Donald J. Ecobichon (consultant), Jeffrey W. Fisher (University of Georgia), David Gaylor (Gaylor & Associates), Ramesh C. Gupta (Murray State University), Rogene Henderson (Lovelace Respiratory Research Institute), Robert MacPhail (U.S. Environmental Protection Agency), and George M. Rusch (Honeywell International).

Although the reviewers listed above have provided many constructive comments and suggestions, they were not asked to endorse the conclusions or recommendations, nor did they see the final draft of the report before its release. The review of this report was overseen by Edward C. Bishop of Parsons Corporation. Appointed by the NRC, he was responsible for making certain that an independent examination of this report was carried out in accordance with institutional procedures and that all review comments were carefully considered. Responsibility for the final content of this report rests entirely with the authoring committee and the institution.

The committee gratefully acknowledges the valuable presentations made by Stephen Channel, Keith R.Vesely, Douglas Somerville, Jeffrey

Gearhart, Sandra Thomson, and Robert Sheridan, all from the Department of Defense. Aida Neel was the program associate, and Cay Butler was the editor. We are grateful to James J. Reisa, director of the Board on Environmental Studies and Toxicology, for his helpful guidance. The committee particularly acknowledges Kulbir Bakshi, project director for the committee, for bringing the report to completion. Finally, we thank all members of the committee for their expertise and dedicated efforts throughout the development of this report.

Gary P. Carlson, *Chair*
Committee on Toxicologic Assessment of Low Level Exposures to Chemical Warfare Agents

Bailus Walker, *Chair*
Committee on Toxicology

Contents

SUMMARY **1**

1 INTRODUCTION **11**
Background, 13
Charge to the Committee, 14
Organization of Report, 14

2 UNDERSTANDING THE PROBLEM OF LOW-LEVEL EXPOSURE TO CHEMICAL WARFARE AGENTS **16**
Background, 18
Toxicologic Studies, 19
Other Animal Studies, 20
Human Exposures, 20
Chemical Warfare Exposure Scenario, 25
Risk Comparisons, 27

3 REVIEW OF THE DEPARTMENT OF DEFENSE MASTER RESEARCH PLAN ON LOW-LEVEL EXPOSURE TO CHEMICAL WARFARE AGENTS **30**
Goals, Objectives, and Aims of the Master Research Plan, 30
Overview of the DOD's Current Research Plan, 31
The Committee's Evaluation of the DOD Research Plan, 32
Overall Summary of the Evaluation of the DOD Research Plan, 62

4 RESEARCH ISSUES RELATED TO RISK ASSESSMENT OF LOW-LEVEL EXPOSURE TO CHEMICAL WARFARE AGENTS 65

Question 1. What Is the Critical Adverse Response to Exposure to Low Levels of CWAs? 65

Question 2. How Much of a Decrement in the Critical Response is Advserse? 66

Question 3. Are There Biomarkers of CWA Exposure That Can Be Measured and Used in Risk Comparisons? 67

Question 4. How Should Exposures in "Work" Conditions Be Considered When Extrapolating Animal Models to Human Experience? 69

Question 5. What is the Relationship Between Exposure Concentration-Duration and Risk? 70

Question 6. How Should Exposure-Duration Risk Be Communicated for CWAs? 73

Question 7. What Are Appropriate Approaches for Evaluating the Risk of Exposure to a CWA Mixture? 74

Question 8. What Information Is Needed to Best Support Extrapolation from Animal Studies to Human Responses? How and When Might a PBPK Model Be Useful for Aiding in Such Extrapolations? 75

Question 9. What Is the Impact of Experimental Designs on Risk Assessments and Subsequent Efforts to Establish Exposure Guidelines for Military Personnel Protection? 78

Question 10. How Does One Account for Variability in Human Response to CWAs, and Do CWA-Susceptible Subpopulations of Humans Exist? 79

Question 11. What Statistical Models Are Appropriate for Concentration-Time Data? 80

Regression Methods for Continuous or Ordered Responses, 85

Experimental Recommendations, 86

The Committee's Recommendations, 86

REFERENCES 89

ABBREVIATIONS 103

Review of the Department of Defense Research Program on Low-Level Exposures to Chemical Warfare Agents

Summary

Historically, research related to chemical warfare agents (CWAs) has focused on life-threatening battlefield effects caused by high-level exposure to CWAs. In contrast, there are limited data on the adverse health effects associated with exposure to low concentrations of CWAs. Concerns have been raised about the potential health effects of longer-term exposure to CWAs at concentrations lower than those needed to produce effects associated with high concentrations. Such concerns have become a priority of the U.S. Department of Defense (DOD). The threat of low-level CWA exposure includes the following:

- Downwind from or at the periphery of a CWA attack or release.
- Entry or reentry into an area after a CWA release.
- Exposures that occur during decontamination operations or from secondary contamination due to incompletely decontaminated material, supplies, and so forth.
- Exposures after planned but inadequate or improper destruction of CWA munitions.

In response to those concerns, the DOD Low-Level Chemical Working Group was formed to develop research programs within the DOD Chemical and Biological Defense Program to understand the adverse health effects of low-level exposure to CWAs, to defend against low-level exposure, to prevent unnecessary duplication of research efforts, and to focus and direct scientific investigations to address operational issues. The DOD Master Research Plan (Research Plan) developed for this effort addresses research on operationally relevant performance decrements and delayed adverse health effects that potentially might be associated with low-level exposure to CWAs.

The stated objective of the Research Plan is to characterize the toxicity of CWAs to enable rational military decision making for issues related to doctrine, training, materiel, leadership, personnel, and facilities. The Research Plan includes consideration of the following: (1) decision making in operational risk management for the range of sensitivities needed for detectors, sensors, and alarms; (2) efficiency needed for individual and collective protection systems; (3) effectiveness of decontamination measures and procedures; (4) restoration of normal military operations; (5) return of previously contaminated materiel to "normal" use; (6) operationally relevant performance decrements and adverse long-term health sequelae; (7) as yet unrecognized outcomes of exposure; and (8) medical diagnostics, prophylaxes, pretreatments, and treatments.

The Research Plan describes DOD's planned research on low-level exposure from Fiscal Year 2002[1] (FY 2002) to FY 2007 that has been or is to be conducted within the existing framework of the DOD Chemical and Biological Defense Program. Every study proposed under the Research Plan is designed to answer one mandatory question: How do data from this work contribute materially toward a quantitative refinement of the human health risk assessment for low-level CWA exposure? The Research Plan is based on three major research thrusts:

- Characterize concentration-time relationships for low-level and longer-time CWA vapor exposures.
- Identify alternative, but physiologically significant, toxicologic end points.
- Conduct appropriate integration studies linking experimental data sets with predictive human health-effect risk assessments.

[1] Prior to 2002, bits and pieces of the research on low-level exposure to CWAs were being done by various departments of the DOD. In 2002, the DOD formalized all the research on low-level exposures in the Department of Technology Office (DTO) of the DOD, and a time line was created to complete various kinds of research by DOD by 2007.

BACKGROUND

In section 247 of the 1999 Defense Authorization Act, the U.S. Congress directed the Secretary of Defense to review and modify DOD policies and doctrines that relate to protecting personnel from low-level exposure to CWAs. In response to that congressional mandate, the Secretary of Defense directed DOD's Nuclear, Chemical, and Biological Defense Program to develop a Research Plan to obtain toxicologic and other relevant data to assess risk to military personnel. According to DOD, the Research Plan is intended to accomplish two objectives. The first is to support operational commanders in the field with information for real-time decision making required to accomplish their missions while not unduly jeopardizing the health and performance capability of their forces. The second is to understand and prevent or reduce operationally relevant performance decrements as well as the potential health consequences of low-level exposure that may not manifest immediately but may become evident months or years after the exposure.

For the Research Plan, CWAs of initial concern are nerve and vesicant agents. Nerve agents include tabun (GA), sarin (GB), soman (GD), cyclosarin (GF), and VX. The vesicant agent of concern is sulfur mustard (HD). Exposure duration and frequencies to be considered are those likely to be experienced by deployed military personnel. Concentrations of interest are those at which no observable adverse health effects (immediate or delayed) are expected for healthy military personnel, as determined by accepted toxicologic tests and standard medical practices. Performance decrements significant to personnel carrying out their military duties are of greater concern than delayed health effects and therefore are assigned a higher priority.

CHARGE TO THE COMMITTEE

The committee's tasks were to review research proposed in the Research Plan, to provide recommendations about that research, and to develop recommendations for additional research as appropriate. The committee's tasks also include providing guidance to DOD on appropriate risk assessment methods for assessing toxicologic risk to military personnel from low-level exposure to CWAs.

THE COMMITTEE'S CONCLUSIONS AND RECOMMENDATIONS

The committee's major conclusions and recommendations concerning the Research Plan are presented below.

- The committee concludes that DOD's master research plan on low-level exposure to CWAs, in general, is well planned and many of the proposed research tasks are likely to provide valuable information in protecting the military personnel from low-level exposure to CWAs and avoiding performance decrements. The Research Plan includes some studies that have some potential to identify delayed adverse health effects, but those studies should be assigned lower priority in the context of DOD's primary objectives. Available information to date do not provide a sound basis for anticipation of delayed adverse health effects following low-level (in particular, short-term) exposure to nerve agents. However, the committee recommends that a small portion of DOD research budget be allocated to some research tasks to rule out the possibility of delayed health effects.
- For DOD, an important task is also to identify the highest concentration of CWA to which an unprotected person can be intermittently or continuously exposed without immediate or delayed health effects. Exposures should range from 1 hour to 1 year, the focus being on acute (short-term, high-level) single exposures and those repeated over 2 weeks.
- Although miosis (pupil contraction) has typically been considered the critical effect and most frequently used indicator of toxicity, questions about other adverse effects at low concentrations of exposure remain. DOD should conduct research to identify whether there are more sensitive toxicity end points from exposure to low concentrations of CWAs.
- If miosis is selected as the critical adverse response, studies must determine what level of pupillary constriction in miosis is operationally relevant. This information is needed for operational risk management and risk-risk comparison, which must consider the range of exposure limits and the probability of adverse outcomes for exceedances of those exposure limits for various levels of toxicity severity and operationally relevant exposure durations. Miosis studies in laboratory animals should include instillation of CWAs in eyes, using appropriate dose ranges, for correlation with cholinesterase inhibition in eyes. Because

ambient light levels confound pupillary effects of CWAs, the committee recommends that the effects of CWAs be determined under known lighting conditions. Furthermore, a rapid and accurate sensor of both miosis and ambient light will be needed to provide information necessary for field commanders to make decisions. There is also a need to review data on decrements in task performance related to changes in pupil size under various conditions of ambient light, including the magnitude and durations of these changes.

- The committee recommends that studies be undertaken to determine whether miosis is the sole cause of operationally relevant performance decrements in humans after low-dose CWA exposure. This research could be done safely in humans with Food and Drug Administration (FDA) approved organophosphate (OP) therapeutic agents and in nonhuman primates with both CWAs and FDA-approved OP agents.
- If cholinesterase levels in the sphincter muscle controlling pupillary constriction do not correlate with functional pupillary changes, then some other mechanism might be responsible for miosis, and a search for alternative macromolecular targets in the ocular tissue might be warranted. The committee recommends that DOD study toxicity end points in addition to cholinesterase inhibition, although tissues (e.g., blood, brain, and eye) should be routinely collected for cholinesterase determinations when animals are killed. Noncholinesterase end points might include fluoride regenerated agent, Na^{+}/K^{+}-ATPase, the Comet assay for DNA damage in lymphocytes, toxicogenomics, and cardiac effects (but not a computer model for cardiac arrythmias).
- One of the tasks of the Research Plan is to study neurobehavioral and cognitive changes at CWA doses below those causing overt toxicity in rodents. According to DOD, neurobehavioral tests, such as the functional observational battery, identify changes in central and peripheral nervous system function beyond those observable as clinical signs of intoxication. The committee understands that animal studies focusing on alterations in behavior are difficult to use when extrapolations must be made across species and to humans. The committee concludes that nonhuman primates are better animal models than rodents for subtle behavior changes in humans. The committee recommends that DOD conduct neurobehavior studies in nonhuman primates. In such animal studies and other studies proposed in the Research Plan, good laboratory practice (GLP) standards must be followed, but no evidence of GLP documentation was presented to the committee. The committee also recommends that DOD identify a behavior, if one exists, that is associated with miosis

and that occurs after systemic (inhalation) or topical (ophthalmic) exposure to CWAs. Review of existing data collected from experiments in humans or nonhuman primates would be of value. Regardless of the particular behavior tests to be used, the committee recommends that consideration be given to evaluating adaptive changes in behavior after repeated low-level exposure to CWAs.

- The committee assigns a high priority to those research tasks that study performance degradation. Performance decrements in humans could be caused by miosis or by more subtle neurophysiologic changes unrelated to miosis. Conducting accurate range-finding studies in a nonhuman primate model will be more useful than extrapolating a maximum tolerated dose from a rodent model or using a dose of CWA that shows an effect in rodents, the method recommended in the Research Plan.
- The committee recommends that data from previous studies on the effects of CWAs on military personnel be reevaluated as potentially useful sources with respect to end points of importance in humans. Between 1958 and 1975, the U.S. Army undertook a human volunteer study of 4,826 subjects to investigate the immediate and long-term effects of various classes of chemicals with warfare potential. The results of that study were perviously reviewed by the National Research Council (NRC) Committee on Toxicology. The committee's first report (NRC 1982) found no evidence to support a finding of adverse long-term or delayed health effects after exposure to nerve agents. However, that report was unable to rule out the possibility that some nerve agents produced long-term adverse health effects in some individuals. In the follow-up NRC study (NRC 1985), 4,000 subjects were sent a health questionnaire. The results indicated that subjects who received nerve agents as a group did not differ from controls who did not receive that chemical treatment, but mortality was lower than expected in the exposed population. DOD also should review the database used to derive acute exposure guideline levels (AEGLs) for nerve agents and sulfur mustard recently published by NRC (2003).
- One of the major tasks of the Research Plan is to develop a method to generate consistent vapor atmospheres for GB across a range of relevant concentrations and exposure durations and relate fairly robust historical vapor generation data to modern, validated chamber exposure methodology. The committee is aware that conducting experiments with potent, volatile CWAs having very steep dose-response curves is exceedingly difficult. The committee concludes that DOD's proposed research to overcome technical challenges to generate consistent vapor atmos-

pheres for CWAs and the deployment of sampling and analytical systems is appropriate, and the committee recommends that such research be continued.

- The military commander or decision maker must have the ability to know if a CWA is present in the ambient air, at what concentration, and whether this concentration can induce adverse operational impairments (performance decrements) or adverse delayed health effects in exposed personnel. However, there are limitations in the detection and measurement of CWAs. The committee recommends that DOD strive to develop more sensitive monitors capable of detecting concentrations of CWAs that can cause performance decrements or potential delayed health effects in an operational setting. The committee also recommends that the degree of detection sensitivity required should be driven by an understanding of CWA toxicology—that is, field-operation detector sensitivity that can identify CWA concentrations that are expected to result in operationally relevant performance decrements or immediate health effects.
- The methodology developed for deriving AEGLs is very pertinent. The generalization of Haber's law to adverse effects related to $C^n t$ (ten Berge method) is a valuable contribution. The committee recommends that DOD utilize information and techniques developed for deriving AEGLs.
- In a deployed military setting, being too protective can be as lethal as being insufficiently protected. Protective equipment not only interferes with an individual's ability to fight, it can also cause significant heat stress. Full protective gear—donning mission-oriented protective posture, level 4 (MOPP4)—can restrict movement and vision and causes dehydration and hyperthermia when used in hot climates. It is critical that operational doctrine not require maximal physically protective measures at exposure concentrations significantly below those likely to produce casualties or long-term disabilities. Therefore, the committee recommends that science-based exposure standards (human toxicity estimates) be determined as accurately as possible and that appropriate toxicologic data be developed to minimize the uncertainty around those values.
- One goal of the Research Plan is to develop biomarkers of CWA exposure and to develop consistent measures directly related to the absorbed dose and physiologic effectiveness of a given chemical agent. The goal is also to demonstrate that the dose-metric profile for exposures in animal models enables direct extrapolation to human physiology.

Proposed biomarkers and dose metrics include alkyl phosphonates, regenerated nerve agent, acetylcholinesterase activity, and butyrylcholinesterase activity. The committee concludes that the proposed research on biomarkers is appropriate and recommends that such studies focus on effects from low-level exposure, paying particular attention to fluoride regeneration of the agent.

- One of the tasks of the Research Plan is to characterize the concentration-time relationship (dose-response) for low-level and longer-time CWA exposures (primarily for miosis and mortality end points). The committee recommends that relevant routes of exposure (e.g., inhalation and local exposure of the eye) be specifically evaluated. The setting of exposure limits for CWAs usually involves extrapolation to concentrations or durations of interest by invoking Haber's law—adverse response is related to the product of agent concentration and exposure duration—or the approach of ten Berge (ten Berge et al. 1986), a generalization of Haber's law stating that adverse response is related to $C^n t$. That expression weighs (1) concentration more than time (when $n > 1$), (2) concentration less than time (when $n < 1$), and (3) concentration and time equally (when $n = 1$). The committee recommends that Haber's law not be applied in the absence of data, because naive applications of Haber's law can lead to erroneous risk estimates. Physiologically based pharmacokinetic modeling can be used to improve the evaluation of concentration-duration-response relationships. Mechanistic considerations also would be informative regarding the conditions and dose metrics for which Haber's law is and is not applicable. In the absence of such mechanistic information, the committee recommends that studies be conducted to test and validate duration-exposure models.
- The committee recommends that the DOD develop and apply appropriate statistical models that include concentration of the agent and duration of exposure as predictor variables along with important covariates that allow for testing various extrapolation methods (e.g., Haber's law and ten Berge's law). To ensure that studies include sufficient sample size, statistical principles of design should be used. Exposure concentrations, durations, and routes of exposure should be selected to include realistic scenarios. To facilitate evaluation of the concentration-time relationship, the committee recommends that exposure-response studies be conducted with a minimum of nine concentration-time treatments (three exposure concentrations crossed with three exposure durations).

- To carry out the operational risk management concept, the risk assessor will require robust information on response probabilities from the experimental study. For CWAs, for each critical response (e.g., miosis), the committee recommends that access to recently generated experimental data sets and contact with investigators be sought to develop response probabilities needed in operational risk assessment beyond single values, such as ECt_{50} (the concentration and time that causes an effect in 50% of subjects).
- Field commanders ultimately must be able to use the information generated by this Research Plan to make decisions. The committee recommends that the DOD risk assessors confer with operations personnel to determine the nature and form of information that would be most useful to field commanders. The committee also recommends consultation with risk communication specialists to further assist with this task.
- One of the tasks of the Research Plan is to identify polymorphisms in human blood esterases and other enzymes or genes to identify susceptible subpopulations. Some proportion of the U.S. population might have genetically determined low levels of plasma cholinesterase and thus unusual susceptibility to some anticholinesterase compounds. Several studies indicate that plasma and red-blood-cell cholinesterase activity is significantly lower in women than in men. Therefore, women might be at a greater risk from systemic exposure to OP compounds. Other enzymes, such as paraoxonase and carboxylesterase, involved in nerve agent toxicity also show polymorphisms, which might result in greater susceptibility in personnel with particular genotypes. This research might be helpful in protecting such individuals using proper risk management strategies.

OVERALL EVALUATION OF THE RESEARCH PLAN

The committee recognizes that a considerable amount of research has been done, and much information is available on the acute and subchronic toxicities (delayed effects) of nerve agents and sulfur mustard. Genetic testing, neurotoxicity testing, metabolic studies, and other research studies have been done. The committee recommends that such studies not be repeated. The committee recommends that the Research Plan not attempt to fill in all the data gaps—that is, investigation in numerous species using multiple dosages by various routes of administration. The time, money, and effort could be better used in focusing on the most important and promising animal models and toxicity end points. The operational relevance of the research in terms of relevant durations of exposure and CWA concentrations must be considered in establishing research priorities.

To obtain the information most valuable in protecting military personnel from operationally significant performance decrements or potential delayed adverse health effects after short-term exposure to low concnetrations of CWAs, DOD should ensure that the total database from previous human and animal studies has been fully examined to fill data gaps. These studies include human studies, nonhuman primate studies, toxicokinetic studies, and the studies used to derive NRC's AEGLs.

1

Introduction[1]

Historically, research related to chemical warfare agents (CWAs) has focused on the significant battlefield threat. This has meant exposures to certain CWAs, such as organophosphate nerve agents and sulfur mustard, at concentrations high enough to produce immediate health impacts. Therefore, there are limited data on effects associated with exposure to lower concentrations of CWAs. On the basis of experiences during and after Operation Desert Storm and with increased emphasis on force health protection, concerns related to the potential health effects resulting from exposure to CWAs at lower concentrations and for longer durations than those needed to produce frank effects at high concentrations became a higher priority for the U.S. Department of Defense (DOD). The threat of low-level CWA exposures includes the following:

- Downwind from or on the periphery of a CWA attack or CWA release.
- Entry into an area after a CWA.
- Exposure to CWA during decontamination operations or from partially decontaminated materiel, supplies, or surfaces.
- Exposure after deliberate (but ineffective) destruction of CWA munitions.

---

[1] Some of the background information in this chapter is from a previous report (DOD 2003).

The DOD Low-Level Chemical Working Group was formed to develop research programs within the DOD Chemical and Biological Defense Program to understand the health effects of exposure to low-level CWAs, to defend against such exposure, to prevent unnecessary duplication of research efforts, and to focus and direct scientific investigations to address operational issues. The DOD Master Research Plan (Research Plan) developed for this research effort addresses research on operationally relevant performance decrements and delayed adverse health effects that potentially may be associated with low-level exposures to CWAs.

The objective of the Research Plan is to fully characterize the toxicity of CWAs to enable rational military decision making for issues related to doctrine, organization, training, materiel, leadership, personnel and facilities; that plan also addresses research on operationally relevant performance decrements. The Research Plan includes, but is not limited to, scientifically credible data necessary to answer questions about decision making in operational risk management; the range of sensitivities needed for detectors, sensors, and alarms; the efficiency needed for individual and collective protection systems; the effectiveness of decontamination measures and procedures; restoration of normal military operations; return of previously contaminated materiel to "normal" use; adverse long-term health sequelae; as yet unrecognized outcomes of exposure; and medical diagnostics, prophylaxes, pretreatments, and treatments (DOD 2003). The purpose of the DOD Research Plan is not to outline a research program to investigate Gulf War illnesses.

This Research Plan describes DOD's planned research, from Fiscal Year 2002 (FY2002) to FY2007, to fill gaps in the toxicologic database for CWAs at low levels of exposure. Research to support the Research Plan objectives is planned within the existing framework of the DOD Chemical and Biological Defense Program. No separate basic research program, organization, or management structure is proposed. The fundamental goal of the DOD research program is to establish science-based exposure standards for military personnel. Every study proposed under that program is designed to answer one mandatory question: How do data from this work contribute materially toward a quantitative refinement of the human health risk assessment for low-level CWA exposures? Therefore, the Research Plan is based on the following three major thrusts:

- Characterize concentration-time relationships for low-level, longer-time CWA vapor exposures.

- Identify alternative, but physiologically significant, toxicologic end points.
- Conduct appropriate integration studies linking experimental data sets with predictive human health effects assessments (DOD 2003).

BACKGROUND

Approximately 700,000 U.S. troops were deployed in the Persian Gulf War. The veterans who served in the Gulf War potentially were exposed to a wide range of chemical agents, including CWAs (e.g., sarin), pesticides, paints, solvents, petroleum fuels and their combustion products, smoke from oil-well fires, and a host of other environmental agents in addition to psychological and physiological stress. Although most men and women who served in that war returned to normal activities, a large number of veterans have reported a range of unexplained symptoms, including fatigue, muscle and joint pain, loss of concentration, forgetfulness, headache, and rash. Many of these Gulf War veterans and some health scientists believe the symptoms are related to the veterans' exposures to CWAs. In section 247 of the 1999 Defense Authorization Act (PL 105-261), Congress directed the Secretary of Defense to review and modify DOD policies and doctrines that relate to protecting personnel from low-level exposure to CWAs.

In response to that congressional mandate, the Secretary of Defense directed DOD's Nuclear, Chemical, and Biological Defense Program to develop a research plan to obtain toxicologic and other data to assess health risk to military personnel. The data obtained from the proposed research would provide information to the Secretary of Defense to reassess policies and doctrines related to low-level exposure to CWAs. According to DOD (2003), the research is intended to accomplish two objectives. The first is to support operational commanders in the field with information for real-time decision making required to accomplish their missions while not unduly jeopardizing the health and performance capability of their forces. The second is to understand and prevent or reduce the potential health consequences of low-level exposure, which might not manifest immediately but could become evident months or years after the exposure.

There is considerable information on acute health effects of short-term, high-level exposures to CWAs, such as nerve agents and vesicants, and there are some data on acute effects of low-level exposure in humans. However, there is currently little information on delayed or chronic effects of low-level exposure to CWAs because before the Gulf War, CWA research focused on lethality and incapacitating effects. Because of interest in the unexplained symptoms among Gulf War veterans and uncertainty surrounding their links with exposures to multiple chemicals, including CWAs and other environmental agents, there has been an increase in research funding by DOD and the Department of Veterans Affairs on potential adverse health effects of low-level exposure to CWAs.

For the DOD Research Plan, CWAs of initial concern are nerve and vesicant agents. Nerve agents include tabun (GA), sarin (GB), soman (GD), cyclosarin (GF), and VX; the vesicant studied is sulfur mustard (HD). Other agents, such as tear gas and hydrogen cyanide, are also of interest to DOD, but they are assigned lower priority. Exposure duration and frequencies to be considered are those likely to be experienced by deployed military personnel. Concentrations of concern are those at and below which no observable adverse health effects (immediate or delayed) are expected for healthy military personnel, using accepted toxicologic tests and standard medical practices. Performance decrements significant to personnel carrying out their military duties are of greater concern than delayed health effects and therefore are assigned a higher priority.

CHARGE TO THE COMMITTEE

The committee's tasks are to review research proposed in the DOD Research Plan to generate toxicologic and other data to help protect military personnel from low-level exposure to CWAs, to provide recommendations regarding that research, and to develop recommendations for additional research as the committee deems appropriate. The committee also will provide guidance on appropriate risk assessment methods for assessing toxicologic risk to military personnel from low-level exposure to CWAs.

ORGANIZATION OF REPORT

Chapter 2 of this report briefly discusses issues related to the toxicity of CWAs. Chapter 3 evaluates the DOD Research Plan on low-level exposure to CWAs and provides conclusions and recommendations for the proposed research; that chapter also identifies gaps in the plan and makes recommendations for further research. Chapter 4 discusses the research issues related to health risk assessment of low-level exposure to CWAs.

2

Understanding the Problem of Low-Level Exposure to Chemical Warfare Agents

Inhalation dose is a function of both concentration and duration of exposure. An exposure dose is considered to be low level if it is below that which results in an immediate observable adverse health effect or operationally relevant performance decrements in healthy U.S. Department of Defense (DOD) personnel exposed to the agent (DOD 2003). The DOD's Master Research Plan (Research Plan) focuses primarily on the latter. While various exposure durations are described in the Research Plan, DOD has stated that a temporary or brief, one-time or continuous exposure lasting from minutes to several hours is the scenario most likely to occur and it is of particular importance in the planning of research.

Chemical warfare agents (CWAs) of primary concern in development of the Research Plan are nerve agents (tabun [GA], sarin [GB], soman [GD], cyclosarin [GF], and VX) and the vesicating agent sulfur mustard (HD); however, nerve agents are of immediate concern to DOD. Throughout the Research Plan, the effects of miosis are referred to as the operationally relevant performance decrements of primary importance. The Research Plan does not point to any known potential delayed adverse health effect associated with low-level exposure to CWAs.

As a result of the high prevalence of unexplained illnesses among veterans of Operation Desert Shield/Storm, a presidential advisory committee (PAC) was formed to address the concerns of veterans groups and the public. The PAC examined the potential health outcomes as they related to selected Gulf War risk factors—for example, exposures to chemical and biological weapons, depleted uranium, infectious diseases, pyridostigmine bromide and so forth. The PAC closely examined the question of adverse health effects after low-level exposure to nerve agents. Their conclusions related to low-level nerve agent exposure were as follows (Presidential Advisory Committee on Gulf War Veterans' Illnesses 1996):

1. Available scientific evidence does not indicate that long-term, subtle, neuropsychological and neurophysiological effects occur in humans after asymptomatic exposure to nerve agents.
2. There are minimal human and animal research data on low-level exposure to nerve agents.
3. DOD should support additional research on the long-term health effects of low-level exposure to nerve agents.

Another reason for DOD to pursue investigations of low-level CWA exposure is based on the military's need to ensure that current doctrine, materiel, and training are adequate to protect soldiers from the effects of exposure to low levels of nerve agents (GAO 1998). While the Research Plan being considered by the committee also may provide some answers to the concerns of the PAC on Gulf War Veterans Health, the Research Plan, and thus the charge to the committee, is directly related to the issue of current military operational doctrine. The questions to be answered by the research accomplished under the Research Plan address operational and delayed adverse health effects concerns of the military and are not designed to develop occupational exposure criteria for the general population. The directed research in the Research Plan should address the development of best estimates of concentration and duration of nerve agents causing mild human incapacitation. One of the issues for the committee was framed by DOD at one of its presentations in terms of the question, When can warfighters safely remove their protective masks without suffering significant performance decrements (miosis) caused by low-level exposure to nerve agents in the environment? The committee assumes the same information would be useful in determining when it is appropriate to put protective masks on when nerve agents are detected. It was stressed to the committee that the focus of the research effort was operational effectiveness as opposed to force health protection. However, it also was stated in the Research Plan that the potential long-term health effects from acute exposure to low levels of nerve agents are to be a consideration for the committee; likewise, the effects of repeated exposures to low doses of nerve agents are in the committee's charge. Therefore, a brief review of the literature on this topic is important in understanding the scope of the problem, and it provides a basis for advancement in knowledge.

BACKGROUND

Four groups of synthetic compounds constitute CWAs: nerve agents (the primary focus of this discussion), vesicants, cyanide, and pulmonary agents. These agents were designed to kill or incapacitate enemy forces, disrupt military operations, and deny terrain to the adversary (Sidell 1992; Chemical Casualty Care Office 1995; Sidell 1997). However, in unscrupulous hands some have been shown to be effective weapons of terror (Yokoyama et al. 1998a,b).

CWAs are classified as persistent or nonpersistent. The former include the vesicants and the nerve agent VX. Nonpersistent agents are volatile and do not remain in an open environment for more than a few hours. Among these are phosgene, cyanide, and the G series of nerve agents—GA, GB, GD, and GF. Toxicity follows exposure to CWAs dispersed as solids, liquids, aerosols, or vapor. Most CWAs were designed to be volatile and nonpersistent and are encountered as vapor or gas. The persistence of the agent depends on factors such as temperature, pressure, and wind speed. Thus, for some of the nerve agents—such as GA, GB, GD, and GF, as well as cyanide, phosgene, and chlorine—the primary route of intoxication is through the respiratory tract (Chemical Casualty Care Office 1995; Sidell 1997). The nerve agents VX and thickened GD and the vesicant agent HD are three of the most persistent CWAs; they pose a threat from dermal absorption of liquids or droplets and can pose vapor hazards as well.

The nerve agents are organophosphorus compounds and their toxicity is primarily related to inhibition of the enzyme acetylcholinesterase (AChE). The rate constants for inhibition of AChE by nerve agents are several orders of magnitude greater than some of the common organophosphates (OPs), such as diisopropyl fluorophosphate (DFP), paraoxon, and methyl paraoxon (Gray and Dawson 1987). Toxicity of the nerve agents is both concentration and time dependent, with acute effects of nerve agents being elicited at very low vapor concentration-duration combinations causing mild symptoms, such as miosis, rhinorrhea, and bronchospasm (Sidell 1974; Chemical Casualty Care Office 1995). Exposure of skin to small-to-moderate amounts of liquid nerve agent causes localized sweating, muscle fasciculation, nausea, vomiting, and lethargy (Sidell 1974; Chemical Casualty Care Office 1995). Large doses of vapor or liquid cause convulsions, loss of consciousness, apnea, paralysis, and death (Sidell 1974; Sidell 1992; Chemical Casualty Care Office 1995). Additionally, after both vapor and liquid agent exposure, there are central nervous system (CNS) effects that vary in intensity and duration. After mild-to-moderate exposure to nerve agent, there may be transient signs such as forgetfulness, inability to concentrate, insomnia, impaired judgment, nightmares, irritability, and depression (Sidell 1974, 1992, 1997).

Most research efforts during the past two decades have focused on developing new antidotal interventions (enzyme reactivators and cholinolytics), pretreatments (pyridostigmine bromide), and preventive measures for nerve agent, seizure-induced brain injury (McDonough and Shih 1997). Several recent comprehensive reviews describing the pharmacology of, and general treatment principles for, the major nerve agents have been prepared by Sidell (1997) and Spencer et al. (2000). Numerous recent comprehensive reviews of the health effects of low-level exposure to nerve agents are provided by Sidell and Hurst (1997), Romano et al. (2001), Brown and Brix (1998), Ray (1998), and Moore (1998).

TOXICOLOGIC STUDIES

For nerve agents, extensive toxicologic studies, including sensitive screening methods, have been conducted in various animal models.

The G agents have been screened for mutagenicity or clastogenicity by using in vitro and in vivo assays (Goldman et al. 1987). GB and GD have been found not to be mutagenic. GA was found to be weakly mutagenic in three different assays but not teratogenic in rabbits and rats (Bucci et al. 1992a,b).

Evidence of neurotoxicity and neuropathology has been sought in a number of studies using nerve agents. Wide-ranging doses of nerve agents were used in 11 studies; no neuropathy or neurotoxic esterase inhibition was seen with VX or with G agents at lethal doses in rats and rabbits treated with atropine and pralidoxime chloride to ensure survival (LaBorde and Bates 1986). According to Sidell and Hurst (1997), "the syndrome (delayed neurotoxicity) has not been noted in the handful of humans severely exposed to nerve agents or in the hundreds of humans with mild to moderate effects from nerve agents." However, more recently, a number of studies have been reviewed that indicate exposure to nerve agents under normal and stressful conditions can inhibit neurotoxic esterase (Somani and Husain 2001).

OTHER ANIMAL STUDIES

Because high doses generally are selected by investigators to elicit observable effects, there are relatively few studies on the effect of low-level exposure to CWAs. Studies have been performed with long-term exposure to symptomatic nonlethal doses of nerve agents (G agents and VX). With the exception of one study using GA in rats, there were no persistent changes in histopathology, hematology, clinical chemistry, or other biochemical parameters (Shih et al. 1994). It should be noted that neuropathology observed following exposure to OP nerve agents is typically thought to be the result of seizures and convulsions following high-dose exposure; thus, single or repeated low-level exposure would not be expected to lead to neuropathological changes.

Attenuation of hormonal responses to physiologic or pharmacologic challenge was observed in a single study 2 weeks after an acute symptomatic dose of GD, possibly attributable to suppression of diurnal hormonal cycles (Kant et al. 1991).

In another study, rhesus monkeys were implanted with cortical and depth electrodes and injected with agent GB in one of two dosage schedules: a single high dose (5 micrograms per kilograms [μg/kg] of body weight, intravenously) that elicited seizures or low doses (1 μg/kg, intramuscularly) once per week for 10 weeks that caused no clinical effects (Burchfiel et al. 1976). Electroencephalograms (EEGs) were recorded before exposure, 24 hours after exposure, and 1 year after exposure. The animals from both dosage schedules had increases in high-frequency beta activity but were otherwise healthy. No long-term behavioral effects were noted.

HUMAN EXPOSURES

For ethical and practical reasons, it is not possible to perform the types of human research studies that would directly answer the questions of concern regarding CWAs over a wide range of exposures. The only human data likely to be available are the results of historical studies (of somewhat limited value for deriving dose-response relationships because analytical and clinical methods were inferior to current methods) or natural experiments (accidental or malicious releases), which generally lack precise estimates of exposure. Suggested below are several ways the value of existing (and future) human data may be optimized.

Accidental Human Exposures

A source of potentially valuable data could be obtained from epidemiologic studies of pesticide-exposed workers, particularly those in underdeveloped countries where personal protection and exposure controls are less sophisticated. While recognizing that those agents do not share all the specific toxicologic properties of the nerve agents, there is sufficient similarity of toxicologic mechanisms that such studies would provide numerous clinical observations that could usefully serve as a means to validate the relevance of animal findings.

Over the past 50 years, hundreds of industrial and laboratory workers have been accidentally exposed to both asymptomatic and symptomatic levels of nerve agents. The strongest evidence in humans of a possible long-term effect of exposure to nerve agents is from studies reporting changes in EEGs. No single study or set of studies exactly addresses the acute and long-term changes in EEG activity produced by nerve agents. Also, virtually all the animal EEG studies of nerve agent exposure have focused on the effects of high-dose exposure and the mechanisms and treatment of more serious toxic effects induced by these exposures.

The first study to suggest long-term EEG effects after nerve agent exposure was by Metcalf and Holmes (1969), who described their findings from a group of workers at a GB production plant. Exposed workers had higher-voltage EEGs with more pronounced alpha rhythm and bursts of slow waves during drowsiness. The individuals also had a high incidence of "narcoleptic" sleep patterns, presumably corresponding to early REM (rapid eye movement) sleep.

Yanno and Musiichuk (1997) published a summary of 209 acute accidental poisonings from Russian nerve agent production facilities. Most, if not all, of the exposures were symptomatic, and a significant proportion of the cohort required hospitalization after their exposure. The resulting adverse health outcomes of the exposures included sleep disturbances and memory loss. CNS symptoms were most persistent in those individuals poisoned by GD.

The widely cited work of Duffy et al. (1979) describes EEG changes found in a group of 77 workers accidentally exposed to GB, showing signs and symptoms of exposure and AChE inhibition greater than 25%. Of the 77, 41 workers had more than three exposures. EEGs from this group were compared with those of a control group from the general population. Subtle but statistically significant increases in EEG beta-band activity over the controls were observed for 1-6 years after exposure in the GB group. Individuals exposed to GB were reported to have more and longer periods of REM sleep than the general population. However, the control group in this study also had changes in their sleep patterns that differed from the general population. The 77 workers exposed to agent GB reported no adverse health effects and no behavioral changes. The significance of the alterations in EEG patterns is uncertain because no behavioral effects can be attributed to them.

Rengstorff (1994) evaluated acute effects of GB on vision in two men accidentally exposed to it. While miosis was evident, no changes in visual acuity were noted. On June 27, 1994, a presumed terrorist attack with GB took place in a residential area of the city of Matsumoto, Japan. About 600 residents and rescue staff were clinically affected; 58 were admitted to hospitals, and seven died (Morita et al. 1995; Yokoyama et al. 1998a,b). Signs and symptoms of exposure included ocular pain, darkness of visual field, nausea, vomiting, headache, rhinorrhea, narrowing of visual fields, sore throat, fatigue, and dyspnea (Nakajima et al. 1998). Two patients had abnormal EEGs (Okudera 2002). Red blood cell AChE inhibition was documented and all examined subjects recovered within 3 months. However, "subclinical miosis" and extremity dysesthesias (reported as "neuropathy") were present in some individuals up to 30 days after exposure. Visual complaints, termed "asthenopia," were present at the outset, more frequent at 4 months, and extant in some individuals 1 year after exposure. A 3-year follow-up study of the exposed population revealed persistence of symptoms among those with lower AChE activity (Nakajima et al. 1999).

Additional information on possible persistent effects after symptomatic exposure to agent GB comes from studies of victims of the Tokyo subway attack on March 20, 1995. Eighteen victims were examined by computerized posturography 6-8 months after the poisoning. Their plasma cholinesterase activities were 13-95 (mean 68.2) international units per liter (IU/L) for females and 19-131 (mean 75.9) IU/L for males.

Romberg quotients for the low-frequency sway in the anterior-posterior direction for females and low-frequency sway and length of sway in the mediolateral direction for males were significantly related to the logarithm of cholinesterase activity. It was suggested that a delayed effect on the vestibulocerebellar system was induced by acute GB poisoning, with females possibly more sensitive than males (Yokoyama et al. 1998). Another follow-up study found P 300 and visual evoked potential (P 100) latencies to be significantly prolonged in GB-exposed persons compared with matched controls (Murata et al. 1997). One subject developed neuropathy with pathological evidence of nerve fiber degeneration at death 15 months after exposure (Himuro et al. 1998).

Studies on Human Volunteers

Between 1958 and 1975, the U.S. Army undertook a human volunteer study to investigate the immediate and short-term effects of various classes of chemicals with warfare potential. The studies were conducted under the auspices of the soldier-volunteer test program of the Army Chemical Center, Aberdeen Proving Ground (formerly Edgewood Arsenal), Maryland. Subjects (n = 4,826) received one or more of 254 chemicals in five classes, including nerve agents. In the early 1960s, 1,406 healthy soldier volunteers, mostly 20-25 years of age, were tested with single or multiple doses of one or more of 15 anticholinesterases, including the OP esters GB (n = 246), GA (n = 26), GD (n = 83), GF (n = 21), VX (n = 740), and DFP (n = 11). Some OP treated subjects were given antidotes. The doses used in these experiments were equal to or less than 1.5 ED_{50} (the dose of agent that causes early signs of incapacitation [miosis] in 50% of exposed individuals). While the quality of medical care and observation fell far short of current standards, data were collected on red blood cell cholinesterase values, symptoms, and signs. Available data suggest that the rate of cholinesterase depression is related to the presence or absence of clinical manifestations. A small number of subjects treated with anticholinesterase chemicals appeared to be unexpectedly sensitive given their unusual reactions during testing. The medical records for two of these subjects were available for review by a National Research Council (NRC) committee (NRC 1982).

The committee recommends that efforts be made to obtain a complete data set for these individuals to determine whether the information included can be used to illuminate the basis for susceptibility.

In 1980, the U.S. Army requested that the Committee on Toxicology of the NRC conduct a study of the possible chronic adverse health effects on the above group of servicemen who had been exposed to chemical agents under experimental conditions. The committee's first report (NRC 1982) found no evi-dence to support a finding of adverse long-term or delayed health effects after exposure to nerve agents. However, this report was unable to rule out the possibility that some anticholinesterase agents produced long-term adverse health effects in some individuals. The report deferred to the outcome of a follow-up morbidity study to shed further light on this issue. In the follow-up study (NRC 1985), a questionnaire was sent to subjects of the earlier studies to assess the current health status of more than 4,000 subjects voluntarily exposed to chemical agents during 1958-1975. The long-term health effects of greatest interest included (1) increased cancer risk and (2) adverse mental, neurologic, hepatic, and reproductive effects. Results indicated that subjects who received nerve agents, as a group, did not differ from controls who received no chemical treatment, but mortality was lower than expected in the exposed population; this reduction in mortality could be due to healthy worker effect. The committee recommends that data from the previous studies on the effects of CWAs on military personnel be reevaluated because the follow-up period was not long enough. The committee recognizes that there were problems with these experiments but nevertheless considers them to be a potentially useful source with respect to end points of importance in humans.

In a follow-up study, Page (2003) conducted a telephone survey of 4,022 military volunteers for a 1955-1975 program of experimental exposures to chemical agents at Edgewood, Maryland. The current health of those exposed to anticholinesterase agents was compared with that of men exposed to no active chemicals (no chemical test) and to two or more other types of chemical agents (other chemical tests). The survey posed questions about general health and about neurological and psychological deficits. There were only two statistically significant differences: volunteers in anticholinesterase agent tests reported fewer attention problems than those in other chemical tests and greater sleep disturbance than those in no chemical tests. In contrast, volunteers who reported exposure to civilian or military chemical agents outside of their participation in the Edgewood program reported many statistically significant adverse neurological and psychological effects, regardless of their experimental exposure. In this study, the health effects of self-reported, nonexperimental exposure, which are subject to recall bias, were greater than the health effects of experimental exposure.

Summary of Studies on Human Volunteers

As published by the U.S. Department of Health and Human Services in the *Federal Register* on March 15, 1988, it was stated that "Questions related to the nerve agents proved relatively easy to resolve. The information bases are fairly complete, and there appears to be little risk either of adverse health effects from long-term exposure to low doses or of delayed health effects from acute exposures." Furthermore, the NRC was confident that its analyses of the Army human volunteer subjects program would have had the power to detect major adverse health consequences had they been present; however, minor or subtle effects could have gone undetected. Until 1991, with the appearance of unexplained illnesses in returning Gulf War veterans, there was little additional debate about the findings.

CHEMICAL WARFARE EXPOSURE SCENARIO

It is important that the research being carried out by DOD be considered in the context of the potential exposure of military personnel. As discussed in the Research Plan, DOD has developed hypothetical scenarios that feature military personnel being exposed to CWAs. The operational conditions in the developed scenarios are intended to address the impact of CWA exposures on military operations and to guide future doctrine and research. The operational assumptions include a fit and healthy military force with knowledge of the threat posed by CWAs and the availability of all defense measures against CWAs. The risk in these scenarios is considered to balance between operational efficiency and personnel safety.

The scenario conditions are as follows:

1. The commander of an infantry unit is directed to capture a well-defended enemy position over a period of 24 hours.
2. The mission is in support of a full-scale contingency, and intelligence indicates a high threat level for the use of CWAs. CWAs may

have been deployed previously on the battlefield with persistent CWAs such as VX, thickened GD, or HD by the enemy to slow a potential advance.

3. Action to take the enemy position might damage existing stores of CWAs and cause their release.

4. The U.S. Forces have available all currently fielded CWA defensive measures.

5. Because of the high ambient temperature (95°F to 100°F), soldiers are directed to wear chemical protective overgarments but not masks and gloves.

For making course-of-action decisions, the following questions need to be addressed:

- Will the possibility of CWAs being deployed by the enemy prevent accomplishment of the mission?
- What is the highest vapor concentration to which an unmasked soldier can be continuously exposed over a 24-hour period without effects that cause a significant decrease in operational effectiveness?
- What is the highest vapor concentration to which an unmasked soldier can be continuously exposed over a 24-hour period without causing chronic or delayed heath effects?
- Should the commander risk the loss of combat efficiency and the likelihood of heat stress by placing troops in full mission-oriented protective posture (MOPP)?
- Should the commander consider the risk of potential long-term health effects from exposure to a low level of CWAs that may be in the environment but cannot be detected with currently fielded technology?

The commander must have evidence that there is a hazard and that the hazard poses a threat. In other words, the commander or decision maker must have the ability to know if a CWA is present and at what concentration and if that concentration can induce adverse operational impairments or adverse delayed health effects in exposed personnel. However, there are limitations in the detection and measurement of CWAs as well as in the validity of human exposure criteria in a deployed military setting.

With regard to detection, a listing of the CWA-detection technologies and the manufacturers of fielded instruments have been compiled in various documents (CBIAC 1995, 1998a, b; O'Hern et al. 1997; NRC 1999).

To maximize the utility of data from accidental or malicious releases, the committee recommends developing more-sensitive chemical sensors to improve exposure data. The sensitivity of the sensors should be sufficient to measure concentrations of these agents at the lowest levels that are considered to result in operational impairment or produce delayed health effects, as discussed in Chapter 4. Enhanced sensors also might be useful on the battlefield, although greater sensitivity of real-time sensors may lead to more frequent false positives, which in turn would risk distracting soldiers from combat-related tasks and overburdening them by requiring them to don physically restrictive personal protective gear.

RISK COMPARISONS

In a deployed military setting, being too protective can be as lethal as underestimating the potency of a CWA. In the above scenario, the CWA is not the only significant health or military threat. Protective equipment not only interferes with the individual's ability to fight, it can cause significant heat stress and produce serious casualties. It is critical that operational doctrine does not require implementing maximum physical protective measures at exposure levels that are significantly below those likely to produce casualties or long-term disabilities. Therefore, human toxicity must be estimated as accurately as possible, and appropriate toxicologic data are required to minimize the uncertainty around those values.

There is a risk of health effects associated with exposure to CWAs. At the most extreme level, death may follow exposure to CWAs. Lesser degrees of impact may inhibit the ability of military personnel to function in combat. Competing with the potential health risks is the risk associated with donning full MOPP. Restrictions in movement and vision coupled with the risk of dehydration when operating in hot climates lower the attractiveness of this option. Strategies are needed for balancing the risk of health effects associated with exposure and with reduced effectiveness when wearing gear. If possible, it might be informative to plot the probability of miosis given CWA exposure versus the probability of severe dehydration or other operational end points for a number of concentration-time combinations. Because these effects differ in severity, an even more important comparison is the probability of an operationally relevant decrement in performance due to CWA exposure versus the probability of a performance decrement resulting from donning MOPP. Data are available on the performance decrement resulting from donning MOPP, but data are needed on the performance decrement (if any) resulting from various degrees of miosis (or whatever the critical effect is, recognizing that this comparison may vary as a function of environmental condition—e.g., ambient temperature).

One of the main objectives of the Research Plan is to support operational risk management decisions with focused research. Army risk management doctrine (Department of the U.S. Army 1998) provides commanders with methods to evaluate and manage the risks posed by operational hazards to the force. Risks are managed by evaluating hazards and implementing operational risk management options during the course of action development. Risk-risk comparisons (balancing exposure to contamination and other risks) are carried out within this existing framework.

This critical premise of the Research Plan is supported by recent recommendations of the NRC report stating the following:

> [T]he establishment of "conservative" estimates of dose-response relations, that is, those designed to err on the side of safety when faced with uncertainty about how to project expected human responses from available data, might not be appropriate for certain military uses. When risks cannot be avoided and decisions are made to accept some risks rather than others, or to bear some risk in furtherance of a more fundamental military objective, it is important to make these trade-off decisions with unbiased estimates of the impacts of various courses of action. In other applications, such as the setting of health-protective exposure standards for application in less severe circumstances, protective estimates might be much more acceptable....[Analyses should be] conducted and...results presented, so that different uses appropriate for different risk-management settings can be made [NRC 2000, pp. 66-67].
>
> Characterization of uncertainty and the limitations of available data are important to all risk analysis, but they might play an especially important role in the analysis of deployment threats, where high-consequence decisions might require taking one risk to avoid others. Risk management ap-

> proaches exist to help make such decisions, but when the risks to be compared are quite uncertain, or uncertain to different degrees, good [characterization] of uncertainty is necessary in order to arrive at sound solutions [NRC 2000, pp. 60-61].

The need for sound science-based risk assessment data was reinforced by the findings summarized in a recent report (NRC 2004). Best exposure guidance estimates are required for making appropriate course-of-action decisions.

In summary, DOD requires accurate and reliable estimates of the effects of low-level CWA exposures on human performance. DOD states that accurate estimations cannot be derived from the universe of existing data on studies with animals and human research subjects and proposes in this Research Plan a multiyear, multimillion dollar research effort to acquire such information.

3

Review of the Department of Defense Master Research Plan on Low-Level Exposure to Chemical Warfare Agents

GOALS, OBJECTIVES, AND AIMS OF THE MASTER RESEARCH PLAN

The overall goal of the U.S. Department of Defense (DOD) Master Research Plan (Research Plan) on low-level exposure to chemical warfare agents (CWAs) is to obtain information that can be used to protect DOD personnel[1] from (1) potential operationally relevant performance decrements and (2) potential delayed adverse health effects from exposures to low levels of CWAs, with initial emphasis on the organophosphorus anticholinesterase nerve agents and sulfur mustard (HD). Other chemicals, such as hydrogen cyanide and tear agents, are given a lower priority. To achieve this goal, the primary objectives of the Research Plan are (1) to obtain the appropriate data for the identification of the most sensitive end point(s) applicable to humans that are indicative of early effects on operationally relevant performance decrements or potential delayed adverse health effects after low-level exposure to anticholinesterase nerve agents and other CWAs, especially during military operations, and (2) to identify strategies for use of the research data for human health risk assessment.

[1] The population of concern includes U.S. military personnel and DOD essential civilians subject to current U.S. deployment policy.

As presented in the Research Plan (Table 5-2, p. 8), the specific objectives are (1) to characterize concentration-time (Ct) relationships for low-level, longer time CWA vapor exposure by conducting in vivo inhalation studies; (2) to identify alternative, but physiologically significant, toxicologic end points resulting from low-level CWA exposures using end points indicative of performance decrements, mechanisms of toxicity, and potential for persistent or delayed health effects (e.g., neurobehavior effects, changes in gene expression); and (3) to conduct appropriate integration studies linking experimental data sets with predictive human health effect assessments using cross-species/cross-route data, toxicokinetic modeling, and biomarkers.

OVERVIEW OF THE DOD'S CURRENT RESEARCH PLAN

The DOD Research Plan states that the nerve agents (G-series agents tabun [GA], sarin [GB], soman [GD], and cyclosarin [GF], and VX) are being, or are going to be, studied alone and in combination with other CWAs after single and multiple exposures by the inhalation route. The highest research priority, as indicated in presentations made by DOD personnel during the committee meetings, is to evaluate the effects of single nerve agents. DOD notes that the dose-response relationships for nerve agent effects studied to date are extremely steep and therefore require a revision of the default uncertainty-factor approach to identify exposure levels that are just under threshold levels. For example, in typical noncarcinogenic risk assessments, a no-observed-adverse-effect level is divided by an uncertainty factor (generally by a factor of 10) to account for variability in sensitivity for intraspecies or interspecies extrapolations. Thus, considering the markedly steep nature of the dose-response curves for effects elicited by these agents, it may be possible to incorporate smaller uncertainty factors (e.g., 3) that would still adequately account for susceptibility differences. The data obtained from the proposed research are to be evaluated for the ability to contribute materially toward quantitative refinement of the human health risk assessments for exposure to low levels of CWAs.

The Research Plan consists of three major research thrusts, with each of the major thrusts subdivided into subthrusts. Subthrusts are further divided into research tasks.

- Major Thrust I: Characterize Ct relationships (concentration-time response) for low-level, longer-time CWA vapor exposures.
- Major Thrust II: Identify alternative, but physiologically significant, toxicologic end points.
- Major Thrust III: Conduct appropriate integration studies linking experimental data sets with predictive human health effect assessments.

The primary question addressed by the committee was, "Do the studies that DOD proposed in its Research Plan match the objectives stated above?" DOD is particularly interested in generating research data that can be used for human health risk assessment and risk management. Risk of adverse effects might ensue from the first attack or from subsequent attacks with CWA munitions. DOD might accept reasonable risk to complete a mission, but information needs to be provided to the commander of the operation to facilitate choosing the most appropriate action(s) (action that reduces risk to personnel but minimizes interference with the military mission). Acceptable risk is expected to be determined by the commander based on the understanding of the situation and mission requirements at that time. Of most concern to DOD currently is a single CWA attack (acute or short-term exposure) and the potential need for mission-oriented protective posture (MOPP), and a deficiency of information on how long such protection may be needed.

THE COMMITTEE'S EVALUATION OF THE DOD RESEARCH PLAN

Major Thrust I: Characterize Ct Relationship (Dose Response) for Low-Level, Longer-Time CWA Vapor Exposures

This major research thrust is subdivided into two subthrusts: (1) Subthrust IA, conduct in vivo inhalation studies to define Ct relationships; and (2) Subthrust IB, improve capabilities to establish and maintain low-level vapor exposure systems for in vivo (animal) inhalation studies. Subthrust IA is subdivided into five tasks (Tasks IA1 to IA5), which are essentially the same. Each of the tasks involves studying the five CWAs (GB, GF, GD, VX, and HD) individually. Experimental studies proposed in the DOD plan for those goals (tasks) either have been conducted for some CWAs or are in progress or planned with mice, rats, guinea pigs, and minipigs. Studies in nonhuman primates are apparently being planned at Edgewood, Maryland and Wright-Patterson Air Force Base (WPAFB). Exposures are primarily by the inhalation route with experimental animals, and it is assumed that results would be qualitatively similar if exposures were by other route. The committee recommends that toxicity end points need to be based on tests other than simple cholinesterase inhibition in blood because this end point has been recognized to correlate poorly with clinical signs. Studies currently proposed by DOD focus on miosis as the most-sensitive operationally significant effect resulting from nerve-agent exposure. Effects on rodent behavior are under investigation as alternative end points indicative of exposure.

Subthrust IA: Conduct In Vivo Inhalation Studies to Define Ct Relationships

This subthrust is subdivided into the following five tasks:

- Task IA1: GB inhalation studies to investigate 50% lethal Ct (LCt_{50}) relationships and miosis (rat and swine).
- Task IA2: GB inhalation studies to investigate LCt_{50} relationships and miosis (rat and swine).
- Task IA3: VX inhalation studies to investigate LCt_{50} relationships and miosis (rat and swine).
- Task IA4: GD inhalation studies to investigate LCt_{50} relationships and miosis (rat and swine).
- Task IA5: HD inhalation studies to investigate nonlethal lung and systemic effects in swine.

These tasks are discussed below.

Task IA1: GB Inhalation Studies to Investigate LCt_{50} Relationships and Miosis (Rat and Swine)

The goal of this research task for agent GB is to demonstrate a dose-response relationship for unequivocal end points of operational significance and to validate or refute the historical database for short-duration exposures. An extension of this goal is to provide new data for exposure durations well beyond previous studies using a well-established animal model for toxicity studies (rat) and an animal model that provides more direct extrapolation to human physiology (swine). According to DOD, this research task will provide a clear description of the Ct-effect profile for GB and predict results in humans based on well-established physiologic extrapolation methods for longer, operationally relevant exposure times (6 hours). DOD research efforts for nerve agents will include studying neurophysiology, biomarker, and dose-metric end points.

Recent studies with GB have been conducted and published by Mioduszewski et al. (2000, 2001, 2002a). Studies of acute lethality and mitogenic response have been reviewed in a National Research Council (NRC) report (2003).

Committee's Evaluation, Conclusions, or Recommendations for Task IA1

As indicated earlier, of note in the data provided by DOD personnel during their presentations at the committee meeting is that miosis is considered an important end point. Much miosis data collected from human studies on GB already exist (NRC 2003, Table 1-7 and Table 1-10, p. 51), and these data should be considered before planning further experiments in animal models to evaluate tissue binding. Evaluation of existing data, identification of data gaps, and sufficient planning would promote appropriate and efficient use of experimental animals. For example, effects of topical exposure on the eye could be discriminated from effects of systemic (inhalation) exposure by conducting experiments with and without eye protection. Also, during experiments examining effects of nerve agents on the eye, it should be possible to examine selected behavioral end points under various ambient light conditions as well. For example, the primate equilibrium platform test, which may serve to measure operationally important performance decrements, has been shown to be sensitive to the effects of anticholinesterase exposures (Hartgraves and Murphy 1992; Murphy et al. 1993; Blick et al. 1994). Therefore, including sophisticated neurobehavioral tests may require the use of nonhuman primates in animal experiments.

The committee further recommends that the research effort also involve studies carried out in mammals other than nonhuman primates. However, the committee recommends that primates be considered the most appropriate species for behavioral studies and for studying the possibility of long-term effects from acute exposures. This recommendation is consistent with the recommendation of the 2003 NRC report. The committee recognizes the issues associated with testing nonhuman primates (e.g., high cost and restrictive laboratory environments), but such testing is recommended in this case. The committee does not make this recommendation lightly but sees such studies as crucial in light of the high precision needed for the estimation of risk estimates for these CWAs, especially since there are competing risks from wearing protective gear.

Regardless of the animals being studied, blood cholinesterase concentrations should be determined routinely, and tissue cholinesterase concentrations from brain and other tissues (e.g., eye) should routinely be collected for determination of acetylcholinesterase (AChE) activity and/or agent regeneration whenever animals are killed. The committee considered the lack of correlation between cholinesterase inhibition and miosis or other low-dose effects of CWAs. Although this may be the case with blood cholinesterase inhibition and changes such as miosis, the correlation between cholinesterase inhibition in the eye and the development of miosis may be high. Experimental studies in which CWAs are instilled in the eyes, with appropriate dose ranges for correlation with cholinesterase inhibition in the eye, may be informative. While blood cholinesterase inhibition may not always correlate with functional changes after CWA exposure, it is a reasonable expectation that miosis and cholinesterase inhibition in the sphincter muscle controlling pupillary constriction will correlate. If they do not, then some other mechanism may be operating, and a search for alternative macromolecular targets in the ocular tissue may be warranted.

The ongoing study evaluating miotic effects of aerosolized CWAs with minipigs was suggested by DOD to be a state-of-the-art approach for evaluating effects of low-level exposures on the putative most-sensitive end point. One issue of concern for animal studies, however, may be the need for restraining the animals to focus the camera on the pupil. It is clear from the DOD Research Plan that although studies to evaluate coexposures to stress along with CWAs are not a high priority, the design of miosis studies in minipigs may have stress as a confounder. A number of reports show that stress alters a variety of physiologic parameters that could be important in responses to CWAs. Thus, researchers should consider measuring plasma corticosteroid concentrations along with other relevant blood measurements during the experimental period. This testing also should be considered for any studies with the primate equilibrium platform, which was found to be a sensitive tool for assessing subtle behavioral effects of GD (Hartgraves and Murphy 1992).

Miosis occurred at 1% of the lethal dose in 50% (LD_{50}) of the minipigs during inhalation studies. Duration of exposure, however, affected the dose-response curve and the calculation of the effective concentration in 50% of the animals (EC_{50}) for this low-dose effect in this species. Information on interspecies relevance of this end point was not provided in the DOD plan or in DOD presentations to the committee, although it was suggested that the EC_{50} for miosis in humans exposed to nerve agents is approximately equivalent to that in minipigs. This is supported by information in the 2003 NRC report.

The pupillary size is reactive to levels of ambient light as well as various concentrations of nerve agents, so the levels of ambient light confound the pupillary effects of CWAs. Therefore, the effects of a CWA should be determined under known lighting conditions to quantify such interactions and to inform field commanders about those lighting conditions under which miosis may be an insensitive (or overly sensitive) toxicologic end point. A rapid and accurate sensor of both miosis and ambient light will be needed to provide information necessary for field commanders to make decisions. It will also be neces-sary to know if the range of decrements in pupil size expected with exposure to low concentrations of CWAs would overlap with the pupil size of unexposed individuals. Existing human and animal data should be of value here (NRC 2003). For operational risk management, there is a need to review data on decrements in task performance with changes in pupil size under various conditions of ambient light, including relative magnitude and relative durations of these changes (such as those induced by the Food and Drug Administration [FDA] approved organophosphate [OP] eye drops—echothiophate). Studies with reversible cholinesterase inhibitors, currently used as therapeutic agents, in people as well as in several animal species including nonhuman primates, could have value. The DOD plan indicates studies would be conducted on various CWAs to evaluate the Ct relationship for each agent in eliciting miosis. If each CWA could be demonstrated to elicit miosis by inhibiting AChE in the iris sphincter muscle (as expected), it might be feasible to model concentration-time relationships by studying organophosphorus-AChE inhibition kinetics in that muscle in vitro and in vivo.

In light of the high precision needed for estimating risk under varying exposure conditions, the committee recommends that studies be undertaken to determine whether miosis is the sole cause of operationally relevant performance decrements in humans after low-dose CWA exposure. This research could be done safely in humans using FDA-approved topically applied anticholinesterase agents and in nonhuman primates using both CWAs and FDA-approved anticholinesterase agents. Furthermore, these studies should be replicated in nonhuman primates using both CWAs and non-CWA agents to determine whether miosis is also the sole cause of performance decrements.

Acute exposure guideline levels (AEGLs) for exposures from 10 minutes to 8 hours are available for the nerve agents GA, GB, GD, GF, and VX (NRC 2003). AEGL-1 (causing nondisabling and reversible effects) values were based on miosis in rats, nonhuman primates, and humans for the G-series compounds. In the NRC (2003) report, the miotogenic response of mammalian eyes was similar across species; therefore, the interspecies uncertainty factor for miosis was considered to be 1. It should be noted that AEGL values are for the general public, including susceptible individuals. Although differences in sensitivity to miosis from direct-acting nerve agent exposures may occur among healthy military personnel, they are not likely to differ substantially from the variability in the general population. It should also be recognized that differences in sensitivity due to differences in metabolic capacity of susceptible subpopulations would likely be of concern only with systemic, rather than local, exposures because differences in biotransformation (or metabolism) would be important only with systemic effects. When establishing AEGL-1 values, NRC (2003) also did not consider other end points that might be more sensitive or more likely to cause operational deficits than miosis.

As noted above, accurate determinations of effect thresholds for CWAs are hindered by the steepness of the dose-response curves. The linearity of these curves at the lower end is still unknown.

The committee concludes that the LCt_{50} studies will probably not provide useful information for low-level exposures because deaths will not occur in animals exposed at low levels. Instead, DOD should focus on effective Ct in 50% of the subjects (ECt_{50}) for other end points that might be relevant to low-level exposures. The committee concludes that dose-duration-response studies provide more robust information than single toxicity end points, such as the LCt_{50} and EC_{50}, for specific exposure durations; therefore, such research is appropriate and should be continued.

Tasks IA2 to IA4

These three tasks are similar to Task IA1 except different CWAs are to be studied. That is, instead of GB, agents studied are GF, GD, and

VX. The committee's conclusions and recommendations generally apply to these tasks also.

Task IA5

HD is a potent vesicant and is also carcinogenic. Considerable toxicity data are available in the literature for HD. DOD proposes to study dose-response for nonlethal lung and/or systemic effects for HD and characterize the dose-response-time-effect profile in swine. DOD has assigned a relatively low priority for this task. The committee recommends that such research be continued with a low priority assigned to it, and thus resources involving these studies should be limited. The committee concludes that DOD's proposed research on HD in swine is appropriate.

Subthrust IB: Improve Capabilities to Establish and Maintain Low-Level Vapor Exposure Systems for In Vivo (Animal) Inhalation Studies

This subthrust is subdivided into the following five tasks:

- Task IB1: GB vapor generation and chamber systems.
- Task IB2: GF vapor generation.
- Task IB3: VX generation and sampling.
- Task IB4: GD vapor generation and chamber systems.
- Task IB5: HD vapor generation and chamber systems.

These tasks are discussed below.

Task IB1: GB Vapor Generation and Chamber Systems

The goal of this research task is to overcome technical challenges to generate consistent vapor atmospheres for GB and relate fairly robust historical generation data to modern, validated chamber-exposure methodology. The DOD goal is also to develop an irrefutable technical method for delivering vapor exposure across a range of relevant concentrations and times. Generation and sampling analytical systems have been developed; comparison of DAAMS (depot area air monitoring system) tubes with Bubbler methodology has been completed by DOD.

Committee's Evaluation, Conclusions, or Recommendations for Task IB1

Conducting experiments with potent, volatile CWAs having very steep dose-response curves is exceedingly difficult. Of critical importance for studies of low-dose effects is a means to ensure a quantifiable, unvarying concentration over the course of data collection necessary for defining specific conditions under which toxicity will increase. The Research Plan recognizes that there is a need for inhalation chambers that provide constant, predictable delivery of CWAs. The Research Plan states that this research will help considerably in determining what constitutes a low-dose effect and whether such an effect will be the same across species.

The proper design for—and safe execution of—inhalation studies with potent CWAs poses a number of challenges. The testing laboratory must create, confine, and define such inhalation exposures with considerable care. Conventional inhalation exposure systems with comparatively large (e.g., 0.25-1.0 cubic meter [m^3]) Hinners-type chamber systems—if operated properly and isolated from the rest of the laboratory—can serve for longer exposure durations. However, studies of short-term exposures (less than 1 hour) might require an alternative approach.

The issue arises from the need to fill and equilibrate the exposure chamber rapidly—ideally, in less than 10% of the total exposure time. Chamber volume, chamber ventilation (flow rate), and exposure duration all need to be balanced, an important requirement for proper execution and evaluation of short-term exposures. If a chamber is operated too slowly, then equilibrium to the targeted exposure concentration is delayed until well into the exposure period. Clearance of the agent or its metabolite postexposure is equally protracted. When evaluating the animals' response (dose-response relationship, the objective of the study), researchers should consider whether eventual peak concentration or time-weighted average concentration best reflects the protracted equilibration and clearance of the exposure system. These considerations bear on the engineering, design, and operation of the exposure systems.

As a consequence, the protective assumption of response linearity from 8 hours to 24 hours was applied by the U.S. Army Center for Health Promotion and Preventive Medicine (USACHPPM) to develop 24-hour military exposure guidelines (MEGs) for nerve agent and HD from the 8-hour AEGLs (i.e., assuming Haber's law that Ct = k for exposure duration greater than 8-24 hours; it should be noted that Haber's law is a special case of the ten Berge expression, when n = 1).

Tasks IB2 to IB4

For DOD's proposed task GF, VX, and GD, the committee's conclusions apply because these agents are structurally similar. However, there are differences in volatility; VX, for example, is harder to volatilize than GB, GF, or GD. Therefore, volatility needs to be considered. VX also adsorbs into chamber surfaces, which makes it difficult to clean chambers between tests with different agents.

Task IB5

The committee concludes that DOD's proposed research in response to the technical challenges in generating consistent vapor atmospheres for HD and the development of sampling and analytical systems is appropriate. DOD proposes to study dose-response for nonlethal lung and/or systemic effects for HD and characterize the dose-response-time-effect profile in swine. DOD has assigned a relatively low priority for this research task. The committee recommends that such research be continued with a low priority assigned to it, and resources involving these studies should be limited.

Major Thrust II: Identify Alternative, but Physiologically Significant, Toxicologic End Points

This major thrust is further divided into four subthrusts, subthrusts IIA to IID.

Subthrust IIA: Identify Acute Pathological Health Effects Resulting from Low-Level CWA Exposures

This subthrust is subdivided into the following five tasks:

- Task IIA1: Establish maximum tolerated doses (MTDs) for parenteral rodent models.
- Task IIA2: Neurobehavioral and cognitive changes in rodents.
- Task IIA3: Neurochemical and immunohistochemical changes in rodents.
- Task IIA4: Cardiomyopathy and cardiac arrhythmia in swine and mice.
- Task IIA5: Cognitive tests in nonhuman primates.

These tasks are discussed below.

Task IIA1: Establish Maximum Tolerated Doses for Parenteral Rodent Models

The goal of this DOD research task is to identify the maximum absorbed dose of each agent that can be tolerated without clinical signs of systemic toxicity and without gross pathologic changes in likely target tissues from a sensitive rodent model. Another goal is to determine a range of doses for nerve agents that could be considered to be low-dose exposures for any relevant duration of exposure up to a maximum of 13 weeks.

Committee's Evaluation, Conclusions, or Recommendations for Task IIA1 The committee reviewed several presentations from DOD personnel that demonstrated that this research task was completed and that the established parenteral MTD was being used in follow-up studies in rodents. The MTD for the nerve agents of interest and the application of these data to follow-up neurobehavioral, pathophysiologic, and toxicogenomic studies is detailed in the Research Plan (pp. 14-15).

Because this work has been completed, it is of no value to make recommendations about whether the research should be conducted. The most important question for the committee is How do data from this work contribute materially toward a quantitative refinement of the human health risk assessment for low-level CWA exposures? (Research Plan, p. 17.) This concern is based on the fact that "cognitive tests will be performed in nonhuman primates using established low-level CWA exposures that show effects in rodents" (Research Plan, Appendix 3, Task IIA5, p. A3-7). The committee's recommendation for primate studies suggests a potential reweighing of this task.

Task IIA2: Neurobehavioral and Cognitive Changes in Rodents

The goal of this DOD research task is to study alterations in behavior that are suspected to occur at CWA doses below those causing clinical toxicity. This hypothesis will be tested in rodents exposed singly and repeatedly to low-level CWA nerve agents. Neurobehavioral tests, such as the functional observational battery, identify changes in central and peripheral nervous system function beyond those observable in clinical signs of intoxication. According to DOD, cognitive performance tests are potentially more-sensitive indicators of subtle performance decrements and include operant conditioning, passive avoidance procedures, and radial arm maze learning tasks. DOD believes data from these studies will identify more subtle alterations in central and peripheral nervous system function due to CWA exposure.

Committee's Evaluation, Conclusions, or Recommendations for Task IIA2 The committee concludes that animal studies focusing on alterations in behaviors are difficult to use when extrapolations must be made across species and to humans. Behavior in a single species can be affected by time of day, environment, age, sex, and so forth. Interspecies extrapolations using behavioral end points will always be difficult, because baseline behavior contributes to whether an effect is seen. Baseline activities even of different species of rodents (rats and mice) are quite different. And although rodents can learn, the importance of their memory compared with that of humans (or nonhuman primates), is unknown, decreasing the significance of rodent behavioral data in risk assessment. These considerations do not negate the potential utility of complimentary neurobehavioral studies in lower mammals; however, nonhuman primates may be better animal models for subtle behavioral changes in humans (Hartgraves and Murphy 1992). The long-term effects reported in some humans exposed to concentrations of OP compounds below those causing toxicity end points indicative of cholinergic poisoning might result from a combination of factors, such as memory, fear, and stress.

Electroencephalogram (EEG) measurements were mentioned during a DOD presentation as a possible sensitive end point indicative of exposure to low concentrations of CWAs, but this has not been correlated with behavioral alterations.

It appears relevant to determine a behavior, if one exists, accompanying miosis that occurs after systemic (inhalation) or topical (ophthalmic) exposure to CWAs. Review of existing data collected from experiments in humans or with nonhuman primates should be valuable. For example, although the primate equilibrium platform test provides a means for detecting subtle behavioral effects of GD in nonhuman primates, it was not reported whether this was correlated with miosis (Hartgraves and Murphy 1992). Other behavioral end points that could be important in the evaluation of alternative toxicologic end points after low-level CWA exposures are deficits in attention and vigilance. Experimental tests—for example, a serial reaction time task (Robbins 2002; Ferraro and Gabriel 2003)—could be incorporated into the testing protocols in humans and both lower and higher mammalian species to evaluate cognitive function that could be pertinent to operational performance. Furthermore, data presented by DOD at a committee meeting suggested that unpublished results indicated visual decrements could occur even in the absence of maximal miosis. Because tests for attention, such as the serial reaction time task, depend on visual discrimination, combined measurements of miosis and visual attention/vigilance or other sensitive behavioral end points could provide important information on performance deficits with low-level CWA exposures.

In addition, DOD should collaborate further with TNO (Nederlandse Organisatie voor Toegepast Natuurwetenschappelijik Onderzoek) in the Netherlands. This institute has studied EEG, visual evoked responses, and miosis in guinea pigs and marmosets and has considerable experience in inhalation exposures for nerve agent exposure as well as analytical capabilities.

Regardless of the behavioral tests to be used, the committee recommends that consideration be given to evaluating possible adaptive changes in behavior after repeated low-level exposures to CWAs and that the connection between measured changes and overt clinical and behavioral effects be evaluated. Repeated exposures to relatively high concentrations of OP compounds are well known to elicit tolerance, at least partially because of adaptive changes in cholinergic receptors. For this context and the proposed DOD studies, focus on pharmacological adaptation along with the evaluation with drug challenges is sufficient. However, the concern of DOD does not focus much on repeated exposures. The dosages of CWAs used in preliminary studies presented, such as repeated exposures at 0.2-0.4 MTDs,[2] were shown to cause substantial AChE in-

---

[2] MTD is defined in the DOD plan as the upper limit of a CWA exposure that can be considered free of clinical signs of toxicity and represents the upper limit

hibition in brain and other tissues and therefore may elicit such cellular tolerance. Furthermore, some OP toxicants, including nerve agents, interact directly with some cholinergic receptor subtypes with high affinity. Thus, changes in cholinergic receptor function might occur with repeated agent exposures due to both indirect and direct interactions. In such cases, adaptation can be evaluated by drug challenge with cholinergic agonists or antagonists. For example, if attention or radial arm maze behavior was being evaluated after repeated nerve agent exposures, a muscarinic antagonist (e.g., scopolamine) or agonist (e.g., carbachol) could be used to evaluate the integrity of cholinergic transmission. Classically, persistent AChE inhibition elicits down regulation of cholinergic receptors. Thus, anticholinergic drug challenge under conditions of cholinergic receptor down regulation can often elicit an exaggerated response. In contrast, cholinergic-agonist challenge under conditions of receptor down regulation often leads to reduced responsiveness. Thus, characterization of dose-related responses to agonists or antagonists can be used to determine whether receptor-mediated adaptation has occurred.

The Navy Environmental Health Center has already deployed such a battery of tests, with human data to validate it. While these behavioral studies were focused on smaller mammalian species, extrapolation to humans is an obvious strength. More communication with and involvement of those investigators in the proposed DOD Research Plan regarding neurobehavioral assessments should be facilitated.

Task IIA3: Neurochemical and Immunohistochemical Changes in Rodents

For this DOD research task, modern techniques will be used to identify changes in brain neurochemistry and cellular-level alterations in metabolism in rodents exposed to low-level chemical warfare nerve agents. According to DOD, data compiled from these studies will provide quantitative estimates of changes in brain chemistry and/or microscopic anatomy in response to chemical warfare nerve agents and will identify doses and durations of exposure required to induce the changes.

---

of low-dose chemical warfare nerve agent for studies.

Committee's Evaluation, Conclusions, or Recommendations for Task IIA3 Data are already available in this area. A portion of this database is summarized in the study matrix references in the Research Plan (Appendix A4). The use of sensitive neurochemical and immunohistochemical methods in the examination of exposed animals is an important component of many of the proposed research efforts. All efforts should be made to include the collection and testing of relevant tissues when studies are designed and to determine the relationship between sensitive measures and more-overt measures of toxicity.

Task IIA4: Cardiomyopathy and Cardiac Arrhythmia (Swine and Mouse)

For this research effort, DOD plans to study the origin of cardiac anomalies seen with chemical warfare nerve agents at low levels of exposures. According to DOD, cardiac failure appears to be a major contributor to nonpulmonary toxicity of these agents in susceptible individuals at low doses. DOD believes conclusions derived from these studies will identify the central or peripheral origin of cardiac arrhythmias after nerve agent exposure as well as the need for, and type of, therapeutic interventions required.

Committee's Evaluations, Conclusions, or Recommendations for Task IIA4 Low-level exposures to CWAs may elicit adverse effects through interaction with other macromolecules in addition to AChE. The research describing interaction of VX with Na^+/K^+-ATPase to elicit cardiotoxicity, along with preliminary findings by DOD investigators that brain Na^+/K^+-ATPase is affected by low-level nerve agent exposures, suggests this macromolecule may be an important additional site of action with some OP toxicants. It is also of interest that the cardiotoxicity of VX was suggested to be due to inhibition of an isoform of Na^+/K^+-ATPase (Robineau et al. 1991). In addition, some OPs bind to muscarinic M2 receptors, including those in the mammalian heart, in a potent manner. For example, recent findings indicate that the OP toxicant chlorpyrifos oxon binds irreversibly to M2 receptors (Bomser and Casida 2001; Howard and Pope 2002). While cardiac muscarinic receptors are prominent in the regulation of cardiac function, ganglionic nicotinic receptors also play a role, and nicotinic receptors have also been reported to be sensitive to direct binding of some OP nerve agents. Thus, there is sufficient preliminary information to warrant examination of additional sites of action with low-level nerve agent exposure that could contribute to cardiomyopathy and cardiac arrhythmia. The committee recommends continuing this research into possible additional sites of action in the heart. The committee also recommends that preliminary experiments could be initiated using cardiac cell lines—for example, the use of cardiomyocytes prior to any in vivo studies to investigate potential alternate mechanisms of cardiac toxicity.

Task IIA5: Cognitive Tests (Nonhuman Primates)

The goal of this DOD research task is to conduct cognitive testing that will be performed in nonhuman primates with established low-level CWA exposures that show effects in rodents. According to DOD, the use of nonhuman primates will permit more sophisticated, computerized testing and will ensure that any observed effects can be extrapolated to human populations.

Committee's Evaluation, Conclusions, or Recommendations for Task IIA5 The committee gives this task a high priority because it will provide the most relevant information useful in establishing the levels of exposure causing operationally relevant performance decrements in humans. Performance decrements in humans could be caused by miosis or by more subtle neurophysiological changes unrelated to miosis. Conducting accurate range-finding studies in the nonhuman primate model will be more useful than extrapolating a MTD from a rodent or using a dose of CWA that shows an effect in rodents, the method recommended in the plan. Studies should be conducted that build on previous efforts that used serial probe recognition after CWA exposure in nonhuman primates (Castro et al. 1994; Myers et al. 2002). The committee supports the DOD plan to conduct neurobehavior studies in nonhuman primates. Neurobehavioral studies using primates are currently on-going at WPAFB. When nonhuman primates are involved in this research, good laboratory practice (GLP) standards must be in place. The requirement for GLP is planned for the execution of all research in the DOD plan; however, no evidence of GLP documentation was presented to the committee.

Subthrust IIB: Determine Mechanism of Action for Relevant End Points Observed at CWA Low-Level Exposures

This subthrust is subdivided into the following two tasks:

- Task IIB1: Toxicogenomic/toxicoproteomics.
- Task IIB2: In vitro electrophysiological and biochemical changes.

These tasks are discussed below.

Task IIB1: Toxicogenomic/Toxicoproteomics

The goal of this DOD research task is to identify changes in gene expression and protein expression induced by exposure to low-level CWAs. According to DOD, the analysis of toxicogenomic data sets will identify doses of CWAs that induce alterations in metabolic homeostasis; it also will establish threshold exposures that have minimal or no effect and will indicate biochemical pathways affected by CWA exposures. For this research task, tissues from animals exposed via inhalation to CWAs are being evaluated to determine whether this end point will correlate with the Ct profile.

Committee's Evaluation, Conclusions, or Recommendations for Task IIB1 The committee believes this type of research might be of some importance when the results suggest possible involvement of noncholinesterase targets. As noted above, if sensitive end points such as miosis can be dissociated from cholinesterase inhibition in the target tissue (in this case, the eye), then these types of screening approaches may be valuable in determining alternative target sites. In this context, a recent report (Chilcott et al. 2003) indicated that mastication after dermal VX exposure occurred before significant reduction in cholinesterase activity and onset of miosis and muscle fasciculations in domestic pigs. Comparative changes in gene and protein expression or differences in phosphorylation of target proteins could provide insight into mechanisms not involving cholinesterase inhibition, especially long-term, latent, and delayed effects. It is important to recognize that these are nontrivial studies and that they need to be related to end points already known. Controls and clear end points are important, especially since effects may occur that are not necessarily directly related to exposure to the test compound.

DOD has included in its Research Plan other experiments that would provide new data of academic interest, but the length of time required to obtain validated end points would make them difficult to adapt for field conditions, especially when acute exposure is of interest. These experiments include the research in toxicogenomics and the research on the binding and recovery of nerve agents from exposed tissues. However, the use of the fluoride-regenerated agent has promise for the further development of dose-response relationships for CWA toxicity at low levels of exposure. As suggested before, inhibition of cholinesterase activity in blood may poorly correlate with physiological changes at low-level exposures. Part of this difficulty is undoubtedly due to the problem in detecting small alterations in the activity of a large set of enzymes (a small signal/noise ratio). The detection of regenerated CWA eliminates this difficulty, with essentially zero background or noise. This inherent difference should provide much higher sensitivity for detecting recovery of agent from tissue samples than for measuring loss of enzyme activity. Use of the fluoride-regenerated agent provides an opportunity to study effects of the agent in human tissues in vitro. Comparative studies on the regeneration of agents in red blood cells from different species, including humans, should be performed—for example, stability conditions, correlation with cholinesterase inhibition, and so forth. The possible effects of dietary fluoride on the regeneration of agents from red blood cells also should be evaluated, and as noted above, the regeneration of agent compared with cholinesterase inhibition in other tissues (e.g., the eye) should be studied to shed light on possible alternative targets (e.g., different binding proteins) involved in low-level effects.

In results DOD presented to the committee, only apoptosis measured by the Comet assay in lymphocytes of guinea pigs exposed to GD and changes in cortical Na^+/K^+-ATPase (increased [3H]ouabain binding) in guinea pigs exposed to GD appeared at exposures less than the MTD. Studies with other nerve agents are lacking. Furthermore, DOD reported that some changes that did occur disappeared rapidly upon cessation of exposure, reducing their usefulness as biomarkers. Results of the Comet assay for apoptosis in lymphocytes were quite robust, however, even at dosages smaller or equal to MTD, potentially representing an additional adverse response to low-level nerve agents. Toxicogenomic studies are only in survey mode. The relevance of these studies in terms of risk assessment of low-level exposures to CWAs are currently unknown. Therefore, these studies should be assigned a low priority.

Task IIB2: In Vitro Electrophysiologic and Biochemical Changes

For this DOD research task, specific alterations in cellular-electrical excitability, synaptic activity, and metabolic state will be studied in brain tissue culture slices and in isolated cell cultures exposed to low concentrations of CWAs. According to DOD, these studies will identify intracellular and neural network mechanisms for low-dose nerve agent effects, differences in agent effects at the cellular level, and pharmacologic interventions required for therapy.

Committee's Evaluation, Conclusions, or Recommendations for Task IIB2 The description of this DOD research task is relatively vague. However, these tests might provide some mechanistic information, although extrapolation of results from single cells and tissue slices to whole animals is extremely difficult. Therefore, the likelihood of generating data that will identify pharmacologic interventions for treatment of individuals with low-level CWA exposures seems remote. Given the overall emphasis on protecting soldiers from operationally relevant performance decrements with low-level CWA exposures, this task seems to be a lesser priority than many others in the DOD plan. The committee therefore recommends that less emphasis be placed and less resources be utilized in performing studies in this task.

Subthrust IIC: Explore the Possibilities of Persistent or Delayed Health Effects Resulting from Low-Level CWA Exposures

Overall, DOD presentations to the committee suggested that this area had lower priority than studies on immediate effects that could interfere with military performance.

The current problem, as presented by DOD, is that determining the lowest dose for effect (either for performance decrement or for adverse health effects) has been difficult. With MTD defined by DOD as the upper limit of exposure without clinical signs, the nonlethal, repeated MTD of nerve agents is approximately 0.2-0.4 LD_{50} per day. Nonlethal effects from exposure to repeated dosages greater than or equal to MTD included significant changes in gait, increased startle response, alterations in passive avoidance behavior, enhancement of mitogen-activated protein (MAP2) activity, EEG changes, messenger RNA (mRNA) alterations, and histopathologic changes on silver staining. Startle response was inconsistent across test compounds.

This research thrust has the following three tasks:

- Task IIC1: Neurobehavioral effects in rodents up to 1 year postexposure.
- Task IIC2: Synaptic receptor density up to 1 year postexposure.
- Task IIC3: Immunohistochemistry and neuron density up to 1 year postexposure.

These tasks are discussed below.

Task IIC1: Neurobehavioral Effects in Rodents up to One Year Postexposure

For this DOD research task, animals exposed either acutely or subacutely will be followed for up to 1 year after exposure, with repeated neurobehavioral testing. According to DOD, low-level CWA expo-sure might produce performance deficits that manifest months after exposure or result only from multiple exposures. Results from animals exposed by inhalation or parenteral routes will be correlated with each CWA profile for the determination of operationally significant effects. According to DOD, findings from these studies will also indicate the potential risk due to low-dose CWA exposures of persistent or delayed-onset neuropathology that could produce long-term disabilities.

Committee's Evaluation, Conclusions, or Recommendations for Task IIC1 The long-term effects of low-level exposures to agent GB (0.5 milligrams per cubic meter [mg/m^3] for 30 minutes) in humans have been studied by Baker and Sedgwick (1996). This exposure caused 60% inhibition of red blood cell AChE activity after 3 hours or 3 days. Human volunteers exhibited miosis and in some cases mild dyspnea. Small changes in single fiber electromyography of the forearm were measured 3 hours and 3 days postexposure and were still detectable at the first follow-up examination 15-30 months postexposure. The authors suggested these electrophysiologic changes "may indicate subclinical onset of nonpolarizing type of neuromuscular block" that is reversible. These findings are subclinical. They are consistent with the research of Senanayake and Karalliedde (1987) with OP insecticides. The committee concludes that such research in laboratory animals by DOD might have limited relevance because, if there are performance decrements, they would be quite subtle and difficult to differentiate from control animals.

Task IIC2: Synaptic Receptor Density up to One Year Postexposure

For this DOD research plan, rodents previously exposed to CWA nerve agents for short-term exposures will be periodically examined after exposure, and brain levels of cholinesterase, muscarinic acetylcholine receptors, and other targeted proteins will be quantitatively determined. According to DOD, alterations in these protein levels and recovery times to normal levels will indicate the rate of recovery of normal brain chemistry after perturbation by low-dose CWA, thus pointing to either improved prognosis with time or lack of recovery.

Committee's Evaluation, Conclusions, or Recommendations for Task IIC2 In general, exposures to either OP CWAs or OP insecticides sufficient to inhibit brain AChE activity substantially for prolonged periods elicit adaptive changes (down regulation) in cholinergic receptors. This down regulation of cholinergic responses in response to AChE inhibition is considered a primary mechanism in the development of tolerance. The adaptive changes in receptor regulation are thought to be elicited through significant increases in synaptic acetylcholine levels after inhibition of AChE.

With very low-level exposures to anticholinesterases, adaptive down regulation would not be anticipated because the level of AChE inhibition likely would be insufficient to alter synaptic acetylcholine levels. However, low-level CWA exposures (repeated MTD exposures) as defined here may be sufficiently high to lead to adaptive receptor changes based on the degree of AChE inhibition elicited. Furthermore, as noted before, some OP anticholinesterases have differential direct effects on cholinergic receptors, in particular the muscarinic M2 subtype. Thus, neurochemical evaluations of specific cholinergic receptor changes after low-level CWA exposures could provide important findings. However, the committee believes that relatively short-term exposures would not be expected to lead to long-term (1-year) alterations in these cholinergic markers. The committee recommends that these studies be considered but that shorter time periods after cessation of dosing be evaluated first before extending to such long-term studies. In addition, the relevance of these changes for extrapolation to whole animals needs to be elucidated. The committee assigns a lower priority for this research task.

Task IIC3: Immunohistochemistry and Neuron Density up to One Year Postexposure

For this DOD research task, the characteristics of brain neurons after exposure will be monitored during and after short-term exposure to chemical warfare nerve agents. Alterations in immunohistochemical staining in existing cells and overall cell number will be monitored after exposure to identify whether any cell populations are selectively affected by the CWA and whether these cells degenerate over time or recover function. According to DOD, such results might indicate changes in brain functions in rodents that are too subtle to detect through behavioral testing.

Committee's Evaluation, Conclusions, or Recommendations for Task IIC3 Henderson et al. (2002) found effects on the muscarinic receptor sites for cholinesterase in the brain only in the combined exposures to GB and heat stress. The committee recommends, based on the investigations of Henderson et al. (2002), that if these tests are conducted, they be done with and without heat stress. The committee, however, concludes that this research task would provide limited information relevant to determining exposure guidelines for low-level exposures.

Subthrust IID: Identify Susceptibility Factors for Low-Level Sensitivities

This research subthrust has the following two tasks:

Task IID1: Identify polymorphisms in human blood esterases.
Task IID2: Computer modeling of cardiac failure to chemical warfare nerve agents.

These tasks are discussed below.

Task IID1: Identify Polymorphisms in Human Blood Esterases

For this DOD research task, differences in levels and subtypes of proteins that bind or degrade OP nerve agents, including chemical warfare nerve agents, will be identified in human blood samples. According to DOD, data sets from these studies will indicate the relative numbers of highly susceptible individuals in the military population. It is possible that testing could identify such individuals.

Committee's Evaluation, Conclusions, or Recommendations for Task IID1 Differences between individuals in blood cholinesterase activity may affect their susceptibility to the toxic effects of CWAs. It has been shown that a small subpopulation of men and women possess genetically determined variants in their plasma cholinesterase resulting in very low activity levels, which therefore may increase their susceptibility to CWAs (reviewed by NRC 2003, p. 156). Several studies reviewed by NRC (2003) indicate that homozygous individuals have plasma cholinesterase activity reduced to less than 25% of the normal value. For heterozygous individuals, mean plasma cholinesterase activity is 64% of normal (range, 28% to 114%). About 3% of individuals may have genetically determined low levels of plasma cholinesterase and therefore may be unusually sensitive to some anticholinesterase compounds. The frequency of the atypical homozygous phenotype is estimated at 0.025% (reviewed by NRC 2003).

Several studies indicate that plasma and red blood cell cholinesterase activity is significantly lower in women than in men (reviewed by NRC 2003). Plasma cholinesterase activity also may be depressed in pregnant women and in people with liver disease, heart disease, allergic conditions, and neoplasms (reviewed by NRC 2003). Such individuals also may be at greater risk from exposure to OP compounds. Although some investigators consider gender differences in plasma cholinesterase activity to be confined to young persons (reviewed by NRC 2003), data are available suggesting that adult females may be more susceptible than males to nerve agents. Yokoyama et al. (1998b, reviewed by NRC 2003) reported vestibulocerebellar effects (increased postural sway) in a small population of patients tested 6-8 months after being exposed to GB during the Tokyo subway terrorist attack. Both female and male patients (nine of each gender) had similar levels of plasma cholinesterase inhibition after the attack, and, in both genders, postural sway was inversely correlated with plasma cholinesterase activity; however, only in females was the increase in sway significantly greater than in controls.

NRC (2003) considered females to be part of the susceptible subpopulation. In the Mioduszewski et al. (2000, 2001, 2002a, 2002b; reviewed by NRC 2003) studies on rats, females were statistically more susceptible than males for the lethality end point. For agent GF, LCt_{50} values generally were lower in adult female rats than in adult male rats (Anthony et al. 2002; reviewed by NRC 2003). The observed increased susceptibility of females was taken into account by applying an intraspecies uncertainty factor for susceptible subpopulations in the estimation of acute exposure guideline levels.

A-esterases (paraoxonase/arylesterase) present in the blood and liver are also capable of hydrolyzing phosphate esters (reviewed by NRC 2003, p. 154). Paraoxonase is also known to be polymorphic in human populations, and individuals express widely different enzyme levels. People expressing certain isomeric forms of the enzyme with low hydrolyzing activity are considered to be more susceptible to OP anticholinesterase poisoning.

Carboxylesterases, another enzyme group capable of binding with certain OP compounds, are present in human blood, monocytes, liver, kidney, and lung. The detoxication potential of carboxylesterases is multifaceted and is an area that requires further experimental characterization.

It should be noted, however, that polymorphisms in blood esterases are important primarily for systemic effects, and they are unlikely to be important for the miotic response that follows topical exposure. If, however, there is concern for susceptibility to systemic effects, DOD's proposed research for this task is appropriate. DOD should also study gene markers for unusual susceptibility to nerve agents and HD. DOD should try to leverage research being done in genomics in other laboratories.

Task IID2: Computer Modeling of Cardiac Failure to Chemical Warfare Nerve Agents

The goal of this DOD research effort is to develop computer models of excitatory activity in networks of cardiac tissue that might be used to identify ionic mechanisms that undergo failure with chemical warfare nerve agent exposures. According to DOD, this research might be used to identify specific ionic channels/biochemical pathways that are highly susceptible targets for chemical warfare nerve agents at low levels, and the findings from this research might lead to the development of therapeutic approaches to treating chemical-warfare-nerve-agent-induced arrhythmias.

Committee's Evaluation, Conclusions, or Recommendations for Task IID2 The committee referred to DOD's concerns for cardiac toxicity of CWA with respect to Task IIA4. In particular, the committee encourages research on "interactions of different CWA with cardiac muscarinic receptors and changes in cardiac function" and also possible low-dose effects on myocardial Na^{+}/K^{+}-ATPase.

By contrast, Task IID2 calls for the use of computer models to identify "ionic mechanisms that undergo failure" due to CWA exposure in order to "identify specific ionic channels/biochemical pathways that are highly susceptible targets" for CWA. The task itself refers to the "modeling of cardiac failure."

The committee has concerns about this task. It is not clear whether the data to be modeled are those to be generated in Task IIA4 or will be obtained elsewhere. It also is not clear that this is intended to address low-level exposures. Task IIA4 is being undertaken to determine whether there are low-level-exposure–related biochemical perturbations leading to cardiac toxicity and dysrhythmias and, if so, to characterize them. Accordingly, Task IID2 seems premature. It is not yet obvious that cardiac effects are important in the context of the principal DOD concerns. It also is not obvious that "cardiac failure" is a relevant end point for low-level exposures.

The committee concludes that Task IID2 has not been adequately described or justified in the Research Plan. It is not clear whether cardiac failure and CWA-induced arrhythmias are appropriate end points for low-level exposures. It is not clear whether cardiac failure refers to myocardial dysfunction or disturbances of cardiac rhythm. The Research Plan refers to agent-induced arrhythmias, but it is not clear whether that actually addresses disordered conduction and/or ectopy, which are potential causes of arrhythmias, or alterations in heart rate (e.g., bradycardia due to muscarinic stimulation, tachycardia due to nicotinic stimulation), which may or may not be sufficient to warrant the term dysrhythmias. The task description does not indicate whether computer modeling will be based on data generated in Task IIA4 or obtained elsewhere; the relationship between Tasks IID2 and IIA4 should be explicitly described. If there are other sources of appropriate data, then their availability might reduce the need for Task IIA4. If not, then Task IID2 cannot be justified until the results of Task IIA4 are available.

Major Thrust III: Conduct Appropriate Integration Studies Linking Experimental Data Sets with Predictive Human Health Effect Assessments

This major research thrust is further divided into two subthrusts: Subthrust IIIA and Subthrust IIIB.

Subthrust IIIA: Cross-Species, Cross-Route (Exposure) Predictive Model Development

This research thrust is subdivided into two tasks. Both of these tasks are considered together.

Task IIIA1: Cross-Species Dosage Comparison and Task IIIA2: Pharmacokinetic Model Development of Agent Distribution Fate

The goal of DOD Task IIIA1 is to demonstrate chemical warfare nerve agent equivalents for two or more experimental animal model systems in predicting well-established end points (e.g., lethality) and compare routes and species dose-response relationships to reconcile historical data sets. The goal is also to deliver a validated basis for predicting dose-effect-time profiles for GB and similar agents to refine future studies and minimize duplication of effort.

The goal of DOD research Task IIIA2 is to develop and extend, if necessary, candidate physiologically based pharmacokinetic (PBPK) models applicable to the human exposure response for CWA and to generate criteria for model evaluation based on the ability of the model to reflect known human and animal exposure data. The goal is to critically assess the model's capability to reflect known human and animal nerve-agent-exposure data and to define any data gaps and, if needed, to determine what parameters and data would be necessary to improve the selected animal model. According to DOD, results from these studies will provide critical operational data to develop guidelines for human activities in a CWA-contaminated environment. Current exposure data for response due to local effects are lacking. DOD also believes the proposed research would use PBPK modeling to overcome some of the problems with extrapolation from species and from differences in exposure conditions.

Committee's Evaluation, Conclusions, or Recommendations for Tasks IIIA1 and IIIA2 In the health-risk-assessment area, the DOD plan appropriately focuses on uncertainties regarding concentration-duration-effect relationships and interspecies extrapolation (Tasks IIIA1 and IIIA2). However, as noted earlier in this chapter and in Chapter 4, many of the interspecies differences are well characterized, and further research should (1) focus on minimizing extrapolation, rather than characterizing it; (2) focus on low-dose effects rather than lethality; and (3) be closely tied to determining the critical effects, since different end points might require different extrapolation approaches. In addition, the use of lethality as an end point for low-level exposures is not a good choice; some other end point such as miosis or ocular irritation should be considered. PBPK models can help to address these uncertainties as well as help address differences in breathing rate between experimental and operational conditions; the development of criteria for evaluating PBPK models is important. One of the key aspects of such models would be to ensure that the model provides information on dose metrics relevant to the critical end points.

Subthrust IIIB: Develop Biomarkers of CWA Exposure and Relevant Toxicologic End Points for Diagnostic or Forensic Purposes

This subthrust is subdivided into the following four tasks:

- Task IIIB1: Correlation of markers of absorbed CWA dose to physiological effects.
- Task IIIB2: Novel protein/genomic markers for CWA exposures.
- Task IIIB3: Identification of cutaneous biomarkers for HD.
- Task IIIB4: Noninvasive neurological tests for CWA exposure (EEG, magnetic resonance imaging [MRI], conduction velocity [CV], single fiber electromyography [SFEMG]).

These tasks are discussed below.

Task IIIB1: Correlation of Markers of Absorbed CWA Dose to Physiological Effects

The goal of this DOD research task is to develop a consistent biomarker in body tissues that is directly related to the absorbed dose of a given chemical agent and its physiological effectiveness. The goal is also to demonstrate that this dose-metric profile for exposures in an animal model provides direct extrapolation to human physiology. The proposed biomarkers/dose metric to be studied includes alkyl phosphonates, regenerated nerve agent, AChE activity, and butyrylcholinesterase activity. Sites monitored would be relevant to the development of pharmacokinetic models that include blood, lungs, kidney, adipose tissue, brain, muscle, and liver. The biomarkers/dose metrics from inhalation and intravenous routes of exposure would be assessed at three levels, ranging from near lethal to a low-level end point such as miosis. Serial blood samples would be drawn during and after exposure and analyzed for each biomarker/dose metric. According to DOD, the data generated from those studies are likely to provide information on which biomarkers/dose metrics should be used for modeling and health risk assessment studies.

Committee's Evaluation, Conclusions, or Recommendations for Task IIIB1 It is vital to obtain reliable indicators of exposure for kinetic analyses. Of the four types of biomarkers to be studied in this task, two of the four are enzymes (AChE, butyrylcholinesterase). An inherent difficulty in using enzyme inhibition as a biomarker of exposure (in particular with low-level exposures) is that the measurement is "loss" of activity. Thus, the signal/noise ratio for low degrees of enzyme inhibition is small. Against the background of substantial activity, a small loss is statistically difficult to demonstrate. This difficulty makes the use of both esterases for this purpose less than ideal. In contrast, the other two biomarkers (alkyl phosphonates and regenerated nerve agent) are not measured against a background of activity (there should be no alkyl phosphonates or nerve agent to be regenerated if exposure has not occurred). This inherently makes these two biomarkers more sensitive indicators of exposure. Thus, the committee recommends making more efforts to correlate either alkyl phosphonates or regenerated nerve agent with physiological effects and making fewer efforts with the enzymes as markers of exposure. The preliminary findings presented by DOD to the committee also demonstrated that the method for detecting regenerated nerve agent had already been successfully used in DOD laboratories and was suitable for multiple CWAs. An added strength of the regenerated nerve agent method is that the agent, instead of one of its metabolites, can be identified from tissue extracts. Furthermore, studies can be used to directly compare regeneration of nerve agent in tissues (e.g., whole blood) from humans and laboratory animals. The committee recommends that future efforts on this task be focused on the regenerated nerve agent method for biomonitoring exposures.

The committee supports the intent of research Task IIIB1, which calls for developing a "consistent marker in body tissues that is directly related to the absorbed dose." On the other hand, it is not clear that such a marker will necessarily correlate with all physiological effects of exposure or at all dose levels, as implied in the task description. Given that underlying DOD concerns are low-level exposures, it is not obvious that assessment of all potential biomarkers at levels of exposure ranging from "near lethal to a low-level end point such as miosis" will be efficient. The committee recommends that studies initially focus on effects from low-level exposure and that particular attention be given to fluoride regeneration of agent. Analytical end points and methods relevant to field concerns and field application should be emphasized. Attention also should be paid to developing markers of absorbed dose for HD.

Task IIIB2: Novel Protein and Genomic Markers of CWA Exposure

The goal of the DOD research task is to evaluate differential enzyme and molecular biological markers for CWA exposure that might supplement cholinesterase as an indicator that low-level exposure has occurred. According to DOD, pattern analysis of multiple biomarkers would determine whether those changes are sufficiently unique to act as diagnostic or forensic markers of CWA exposure.

Committee's Evaluation, Conclusions, or Recommendations for Task IIIB2 In the case of organisms exposed to G agents, the presence of the agent is detectable for only a few hours. Therefore, intact G agents are not good candidates for retrospective detection of exposure. The metabolism of G agents takes place primarily by hydrolysis. In addition to binding to AChE, they bind with the closely related plasma protein butyrylcholinesterase and to carboxylesterase; binding to serum albumin also occurs (Noort et al. 2002).

Polhuijs et al. (1997) developed a method for analyzing phosphorylated binding sites (e.g., butyrylcholinesterase), which is based on reactivation of phosphorylated enzyme with fluoride ions. The OP-inhibited butyrylcholinesterase in human plasma has a half-life of 5-16 days. Another method for detecting exposure to nerve agents involves electrospray tandem mass spectrometric analysis of phosphylated nonapeptides obtained after pepsin digestion of butyrycholinesterase from human serum samples (Fidder et al. 2002).

The DOD research efforts to develop or identify novel protein/genomic markers for CWA exposure are appropriate and might provide some information in identifying exposed people, which would help determine the treatment or management of the exposed personnel. Studies to date, however, suggest that it will be difficult to obtain results that are relevant and useful to DOD. This task would have more utility for HD.

Task IIIB3: Identify Cutaneous Biomarkers for HD

The goal of this DOD research task is to develop antibodies that can recognize DNA or keratin adducts with HD as potential indicators of exposure. HD binds to proteins and DNA to produce adducts that are immunologically distinct. DOD will also explore the utility of proteases and cytokines produced in human keratinocytes in response to exposure to HD as alternatives to existing diagnostic methods. According to DOD, biomarkers of HD exposure identified by those studies might prove to be better diagnostic tools than existing tests of metabolic products of HD.

Committee's Evaluation, Conclusions, or Recommendations for Task IIIB3 HD is a bifunctional alkylating agent that reacts rapidly with nucleophiles. Its metabolites are excreted in urine within 24 hours, yet a significant portion of the absorbed dose persists in the blood for weeks to months, depending on the species. The use of urinary metabolites as biomarkers offers the advantage of a noninvasive test, but its value is limited to very recent exposure, because the urinary metabolites are detectable for only hours to days after exposure. This limits their use for retrospective detection. HD forms many protein adducts by alkylation. In a group of victims exposed to HD, the amino-terminal valine adducts corresponded with approximately 0.9 micromoles (μmol) of HD. These findings were confirmed by immunochemical analysis of DNA adducts in lymphocytes from the same blood samples (reviewed by Noort et al.

2002). A sensitive method to detect protein adducts has been developed that uses liquid chromatography (Noort et al. 1997; Black et al. 1997). Recently, Van der Schans et al. (2002) showed that most of the radioactivity (80%) of radioactive HD was bound to keratin. Van der Schaus et al. (2002) developed a direct detection method for these adducts in stratum corneum of human skin that used immunofluorescence microscopy. This method has the potential to lead to the development of a rapid detection kit that can be applied to the skin (Noort et al. 2002).

The committee recommends that biomarkers for routes other than cutaneous also be developed because there could be noncutaneous exposures.

Task IIIB4: Noninvasive Neurological Tests of CWA Exposure (EEG, MRI, CV, SFEMG)

The goal of this DOD research task is to determine the diagnostic utility of changes in nerve and muscle activity that previously have been shown to correlate with exposure to a CWA in rodents. If those previous findings can be replicated and if the effects are adequately robust, then their use will be considered for population-based studies. According to DOD, such studies will indicate whether changes in EEG, nerve CV, SFEMG, or MRI can serve as reliable diagnostic markers of CWA exposures.

Committee's Evaluation, Conclusions, or Recommendations for Task IIIB4 The noninvasive methods listed above for evaluating potential effects of CWAs may have general utility under some conditions. For example, MRI provides a powerful tool for evaluating morphological changes in tissues (e.g., brain) in the same individual over time. Such measurements may be particularly important in evaluating persistent, delayed responses to CWA exposures. However, because the major goal of DOD research is to determine exposure levels that could lead to operationally relevant performance decrements and delayed health effects and the exposures to be modeled are for low doses, which are likely to produce subtle changes at most, those methods might not provide information of highest priority for the objectives of this research program. The committee recommends that, with the limited resources and time available for completion of the proposed research, less emphasis be placed on continuing the research proposed in this task. If this research is pursued, it should be done in nonhuman primates before suggesting it would be useful for humans. These tests are also difficult to adapt to field conditions.

OVERALL SUMMARY OF THE EVALUATION OF THE DOD RESEARCH PLAN

The DOD Research Plan identifies research needs that include experiments likely to provide data to contribute to the understanding of concentration-time relationships, the identification of biomarkers of exposure, the indicators of susceptibility and variability of response, improved capabilities and methods for assessing the potential to cause delayed adverse health effects, the description of mechanisms of toxicity associated with mixed and combined exposures, relationships between CWAs and other factors, and hazard assessment models. DOD presentations to the committee indicated that the current emphasis is on single exposures to single agents and the immediate effects on operational performance. For DOD, an important task is also to identify the highest level of CWA to which an unprotected person can be intermittently or continuously exposed without immediate or delayed health effects when exposures range from 1 hour to 1 year, focusing on exposures from acute single exposures to those repeated over 2 weeks. The primary objectives of DOD is the concern for the degradation of performance of military personnel from exposures to CWAs, and that concern makes the research on immediate effects more important.

The committee recognizes that a considerable amount of research has been done and much information is available on the acute (short-term, high-level exposures) and subchronic toxicity of nerve agents and HD. Genetic testing, neurotoxicity testing, metabolic studies, and other research studies have been done. The committee recommends that those studies not be repeated. The committee recommends that the Research Plan not attempt to fill in all the data gaps—that is, not investigate every species by various routes at multiple doses. The time, money, and effort could be better used in focusing on the most important and promising animal models and toxicity end points. The operational relevance of the research in terms of duration of exposure and CWA concentrations must be considered in establishing research priorities. The committee makes these recommendations because these studies should have operational relevance and the current interest is for consistent, sensitive detection of adverse responses after low-dose, short-term exposures to CWAs.

The committee recommends that the Research Plan include the development and application of appropriate statistical models for the data being generated. Specifically, statistical models should include concentration of the agent and duration of exposure as predictor variables along with important covariates that would allow for testing various extrapolation methods (e.g., Haber's law, ten Berge's law). Studies should include sufficient sample size at each tested exposure concentration and duration to detect changes that are judged to be biologically significant. To adequately determine the appropriate size of a planned study, statistical principles of design should be used. The following issues must be considered when determining animal numbers in an experimental study: (1) the magnitude of a difference that is considered significant (e.g., amount of pupillary constriction), (2) the magnitude of variation in response that is expected, and (3) the sensitivity/power that is desired for detecting this difference. Exposure concentrations, durations, and routes of exposure should be selected to include scenarios that are as realistic as possible.

To obtain the information most valuable in detecting and protecting military personnel from operationally significant decrements or potential delayed adverse health effects after short-term exposures to low levels of CWAs, DOD should ensure that the total database from previous human and animal studies has been fully examined to fill data gaps. These include specific studies done with human subjects (NRC 1982, 1985), studies in primates (Hartgraves and Murphy 1992), toxicokinetic studies (Benschop and De Jong 2001), and the studies used to derive AEGL-1 values (NRC 2003). Then, studies should be designed to correlate the most likely operationally significant decrement (e.g., miosis under different ambient light conditions after topical and inhalation exposures) with subtle behavioral change(s). When possible, actual tasks of military importance (e.g., visual tracking on radar screens) should be evaluated after topical OP exposure. These studies would best be done in humans (by using FDA approved therapeutic OP agents) or nonhuman primates, although the value of appropriately designed, efficient preliminary experiments in lower mammalian species (e.g., rodents and minipigs) is recognized. Wherever possible, associations also should be made with cholinesterase inhibition and/or reactivation or tissue binding of CWAs, especially in target tissues. Although unlikely to provide immediate results to assist with detection of and protection from CWAs in the field, other studies proposed in the DOD Research Plan are likely to have future value (e.g., toxicogenomics, in vitro cellular and biochemical effects). The committee recommends that some research with potential long-term rewards be continued even in the absence of immediate applications. Such experiments may provide insight into delayed adverse health effects. That research includes studies on the binding of CWAs to and reactivation from target and nontarget tissues, toxicogenomics, and biochemical effects noted at concentrations less than the MTD.

The committee concludes that DOD's Master Research Plan on low-level exposures to CWAs, in general, is well planned and many of the proposed research tasks are likely to provide valuable information to DOD in protecting military personnel from low-level exposures to CWAs, in terms of avoiding performance decrements and delayed health effects. The Research Plan includes some studies that have some potential to identify delayed adverse health effects, but those studies should be assigned lower priority in the context of DOD's primary objectives. Available information to date does not, however, provide a sound basis for anticipation of delayed adverse health effects following low-level (in particular, short-term) exposures to nerve agents. However, the committee recommends that a small proportion of the DOD research budget be allocated to some research tasks to rule out the possibility of delayed health effects.

4

Research Issues Related to Risk Assessment of Low-Level Exposure to Chemical Warfare Agents

As noted earlier, one charge to the committee was to review and consider the U.S. Department of Defense's (DOD's) research plans for evaluating low-level exposures to chemical warfare agents (CWAs), identify data gaps, and make recommendations for further research. The second part of the charge asked the committee for guidance on appropriate risk assessment methods for assessing health risk to military personnel from low-level exposures to CWAs. To address this, the committee took the approach of coming up with a series of questions that must be addressed as part of the risk assessment of low-level exposure to CWAs, as a way to focus research on the most important issues.

QUESTION 1. WHAT IS THE CRITICAL ADVERSE RESPONSE TO EXPOSURE TO LOW LEVELS OF CWAS?

The U.S. Environmental Protection Agency (EPA 2004) defines the critical response as "The first adverse effect, or its known precursor, that occurs to the most sensitive species as the dose rate of an agent increases," and it must be considered whether the most sensitive species is relevant to humans. Other organizations use somewhat different definitions, but identifying the first effect of concern for the population of interest as dose increases is a key step in the development of exposure guidelines. For CWAs, an adverse effect is considered an operationally relevant performance decrement or an adverse health effect.

Throughout the DOD report, low-level exposures to CWAs that may cause miosis have been emphasized (pupillary constriction); that effect occurs earlier and more reliably than other adverse effects. Although miosis typically has been considered the critical effect and most frequently used indicator of toxicity, questions about other adverse effects at low-level exposures remain. As noted in Chapter 3, Hartgraves and Murphy (1992) and Wolfe et al. (1992) observed decrements in equilibrium performance in primates at soman (GD) doses that generally do not produce classic signs of cholinergic toxicity, and Chilcott et al. (2003) found that mastication consistently preceded miosis and salivation in minipigs treated with high percutaneous doses of VX. Therefore, DOD should conduct research to determine whether there are more sensitive toxicity end points than miosis from low-level exposures to CWAs.

QUESTION 2. HOW MUCH OF A DECREMENT IN THE CRITICAL RESPONSE IS ADVERSE?

In the military operational risk management (ORM) context, this means identifying the degree of decrement that is operationally relevant, including the potential for the decrement to result in a field unit to become ineffective for combat or other mission-related activities. This information is currently not available. As noted above, if miosis is selected as the critical response, studies must determine what level of pupillary constriction from miosis is operationally relevant. For example, determining that a specific exposure or dose will cause a 10% decrement in pupil size is not sufficient without knowing whether that decrement is operationally relevant and determining what other operational factors must be considered. For example, a degree of pupillary constriction that represents no decrement in performance for daylight operations may represent a significant impairment for evening or night operations. Under daylight conditions with adequate lighting, a moderate degree of miosis may not be operationally relevant. It is normal for human pupil diameters to decrease by 50% when moving from dim to bright light conditions, such as when one moves from a dark room to outside when the sun is shining brightly. This suggests that different CWA exposure limits may be appropriate for use under different lighting conditions. Furthermore, miosis has been reported to improve visual acuity of presbyopic individuals (people who have lost elasticity of the lens due to advancing age) under well-lit conditions, a finding attributed to the pinhole effect, but miosis would decrease visual acuity in dim light (Sidell 1977). If visual acuity is affected, how much loss of acuity is important and how long does it last? Risk managers generally prefer a more cautious or protective approach when the health effect of interest is permanent or severe. In contrast, implementation of ORM risk-risk comparison requires the development of exposure limits and probabilities of adverse outcomes for exceedances of those limits for each level of severity and operationally relevant exposure duration. The field commander must consider the severity of delayed effects influencing future performance as well as possible immediate effects both from exposure to CWA and from heat stress resulting from wearing protective gear (mission-oriented protective posture [MOPP]).

Although it is desirable that the demarcation between acceptable and adverse be defined by the content specialists (e.g., how much miosis would impact a war fighter's ability to perform his/her mission successfully), for neurobehavioral effects that are measured on a continuous scale, there often is no clear point of demarcation between an acceptable and adverse level, particularly for studies in laboratory animals. In such cases, the distribution of effects in unexposed control animals may be used to establish an abnormal range—for example, above the 95th and 99th percentile. Then, the probability of animals exhibiting responses in the abnormal (not necessarily adverse) range can be estimated as a function of exposure, using the available concentration-duration-response data. Such procedures have been used for behavioral data (Gaylor and Slikker 1990).

QUESTION 3. ARE THERE BIOMARKERS OF CWA EXPOSURE THAT CAN BE MEASURED AND USED IN RISK COMPARISONS?

Biomarkers of exposure and of effect are essential to understanding the relationships between concentration and duration of exposure and the resulting effect. A biomarker of exposure is a xenobiotic substance or metabolite or the product of an interaction between a xenobiotic and a target in the organism (NRC 1989). A biomarker of effect is a measurable biochemical, physiologic, or other alteration within an organism that, depending on magnitude, can be recognized as an established or potential health impairment or disease (NRC 1999). A biomarker of susceptibility is an indicator of an inherent or acquired limitation of an organism's ability to respond to the challenge of exposure to a specific xenobiotic. It can be an intrinsic genetic or other characteristic or a preexisting disease that affects the toxicokinetics of the chemical or the target tissue response to the chemical.

Recent research has used protein binding and fluoride-induced regeneration of the agent as a biomarker of exposure to CWAs. Biomarkers of exposure can be used as forensic biomarkers to aid in the reconstruction of exposure levels and the normalization across different exposure levels to a common dose metric. They can also be used to determine post hoc whether someone was exposed under combat conditions and to determine whether any apparent delayed effects are related to exposure. Such use, however, depends on the stability of the biomarker of interest, because it may be weeks or months before exposed personnel are removed from the field and assayed for possible exposure. Another research question is whether sampling weeks or months after exposure can be used to identify whether someone was exposed, and, if so, to quantify that exposure. Although the uses of biomarkers described in this report may have limited relevance to immediate ORM decisions, it is likely they might be useful for research on postexposure sequelae, for refining interspecies extrapolations, and for exposure reconstruction (the DOD Research Plan only mentions biomarkers briefly).

See discussion of Tasks IID1, IIIB1, and IIIB2 for other issues related to this question.

Decreased blood cholinesterase generally is considered a biomarker of exposure, although some scientists consider it to be sufficiently closely tied to adverse end points that they consider it to be a biomarker of effect. However, as noted earlier, it is not always a reliable indicator of toxicity after exposure to organophosphates (OPs). While decreased brain cholinesterase activity is a more appropriate biomarker of effect, it is obviously not a useful end point that can be studied in the field.

Biomarkers of susceptibility might be useful before exposures to CWAs to evaluate the potential increased risk to susceptible individuals. People with polymorphisms (genetically determined variants) that result in very low plasma cholinesterase activity may be unusually sensitive to systemic effects of some anticholinesterase compounds (Young et al. 1999). Similarly, people with genetic polymorphisms leading to lower activity in the enzymes that metabolize CWAs to inactive forms may have enhanced sensitivity, although the implications of such polymorphisms for determining tissue dose (and therefore risk) depend on the degree to which enzyme activity limits metabolism (Gentry et al. 2002). This discussion is probably most relevant for systemic toxic effects. It may not be relevant for miosis if this response is primarily the result of direct contact with a CWA.

QUESTION 4. HOW SHOULD EXPOSURES IN "WORK" CONDITIONS BE CONSIDERED WHEN EXTRAPOLATING ANIMAL MODELS TO HUMAN EXPERIENCE?

Military personnel are likely to be exposed to CWAs while physically active. Thus, the use of animal models in which exposures are conducted while the animals are physically restrained or inactive may be problematic. In occupational settings, the higher respiration associated with physical activity will result in higher systemic exposures, while restraint-induced stress may lead to increased adrenergic stimulation. Also, depending on the location of deployment, military personnel often operate under severe environmental conditions, such as extremely high or low ambient temperatures. These environmental conditions could affect respiration and uptake of CWAs and may alter the expected degree of impairment associated with exposure to the agent. Experimental design should consider incorporating such factors (e.g., physical exertion with treadmill exercise, increased temperature) as variables where feasible.

Adjustments are also needed in extrapolating from studies in the general population or worker population to military scenarios. Previously, when human volunteer data were used to develop airborne exposure guidelines for G agents in occupational settings (Mioduszewski et al. 1998), it was necessary to extrapolate from the resting respiratory rate associated with the human volunteer studies to that associated with an occupational setting. The extrapolation assumed a human resting respiratory rate of 10 liters (L) per minute and an occupational respiratory rate of 20.8 L per minute over 8 hours. Different minute-volume adjustments would be necessary for deployed military personnel because, depending on the military missions, their activity levels and exposure durations would be different from a typical 8 hour/day worker. Strenuously working individuals will absorb more agent than people in the general population because of higher pulmonary ventilation and higher cardiac output.

Appropriate consideration of activity levels also may depend on the toxicity end point used to derive exposure limits. For example, systemic and respiratory effects are closely tied to the minute volume under many conditions, and ocular effects resulting from topical exposure, such as miosis, may be a function primarily of the exposure concentration and duration of exposure.

QUESTION 5: WHAT IS THE RELATIONSHIP BETWEEN EXPOSURE CONCENTRATION-DURATION AND RISK?

The setting of exposure limits based on dose and duration has been considered in a variety of contexts. For example, Jabarek (1995) presented a collection of potential dose metrics along with an accompanying illustration. Although ease of use has made it a common practice to invoke Haber's law (the assumption that adverse response is related to the product of agent concentration and duration of exposure, or Ct or cumulative exposure metric), Atherley (1985) provided a historical summary and challenge to the routine use of this assumption.

ten Berge et al. (1986) provided a generalization of this relationship that weights concentration more than time (for $n > 1$), stating that adverse response is related to $C^n t$. Using this relationship, they found that, for acute lethality, the exponent (n) was greater than 1 for 19 of 20 tested chemicals and the highest exponent was 3.5. Similar broad analyses of the ten Berge et al. (1986) modification have not been conducted for end points other than lethality. Use of the ten Berge relationship for extrapolating to durations well beyond experimental design is problematic because its applicability for those scenarios has not been studied or validated. Mechanistic considerations can also aid in improving on the ten Berge relationship. Empirical curve-fitting methods can also be used if the data span the time period of interest. Physiologically based pharmacokinetic (PBPK) modeling (see Question 8) can also be used to improve the evaluation of concentration-duration-response relationships.

Another tool that has been used to evaluate concentration-duration relationships, particularly for acute inhalation exposures, is categorical regression. This is a method by which a dose-response model may be fit to data where only severity ratings are available (Hertzberg and Miller 1985; Hertzberg 1989; Guth et al. 1997; reviewed by Haber et al. 2001). For example, a study may note that exposure to 15 milligrams per cubic meter (mg/m^3) for 10 minutes caused miosis, while clinical signs of toxicity occurred after exposure to the same concentration for 1 hour. Thus, an advantage of categorical regression is that it can be used to analyze the concentration-time severity of response relationship in the absence of quantitative data.

Often, acute inhalation data from one study are insufficient to describe the full concentration-duration-response spectrum of interest. However, data on the effects of interest may be available from studies conducted in different species and strains of animals and different concentration-time patterns. Categorical regression can be used to combine these data in the development of acute inhalation exposure limits (NRC 1993; Guth et al. 1997). EPA has developed guidance on use of categorical regression (EPA 2000), and their free software can be used to conduct the modeling (EPA 2003).

Several other methods of combining studies are also available, although there is less experience in applying these methods for toxicological risk assessment. For example, meta-analytic methods are frequently used to combine estimates across independent studies (Hedges and Olkin 1985; Gurevitch and Hedges 1993; Piegorsch 1998; Manly 2001; Guth et al. 1997). These estimates are typically weighted by some estimate of variability/uncertainty associated with each of the potency estimates. Alternatively, analyses considering strain and species as random effects and concentration and duration as fixed effects might be conducted (Diggle et al. 1994; Verbeke and Molenberghs 1997). Risk estimation in the context of these mixed effects models may represent an area of interesting risk assessment research.

The methods developed for acute exposure guideline levels (AEGLs) are very pertinent for exposures to CWAs. The generalizations of Haber's law to effects related to $C^n t$ (ten Berge et al. 1986) is a valuable contribution. The National Research Council (NRC) (2001) recommends the use of this approach for developing AEGLs for CWAs. The NRC (2003) used the ten Berge et al. (1986) approach in developing AEGLs for exposures of 10 minutes to 8 hours. For sarin (GB) (and subsequently tabun [GA], soman [GD], cyclosarin [GF], and VX), they used an exponent of $n = 2$, based on both miosis and lethality data. For sulfur mustard (HD), a value of $n = 1$ was used for the lower-severity values (AEGL-1 and AEGL-2). For AEGL-3, which is designed to protect against life-threatening effects or death, the NRC (2003) used the health-protective approach of using an exponent of $n = 3$ for extrapolating to shorter periods (resulting in a flatter curve in the time-exposure-duration dimension relative to the concentration-dose dimension) and $n = 1$ for AEGL-3 for extrapolating to longer periods.

For military operations, the U.S. Army Center for Health Promotion and Preventive Medicine (USACHPPM) has developed air military exposure guidelines (MEGs) for CWAs (USACHPPM 2002a,b). These MEGs are derived based on existing data and assumptions about critical end points such as miosis and lethality. The findings from research on neurobehavioral and cognitive effects would have great importance on future updates and revisions of the MEGs. As such, coordination and communication of research strategies between DOD risk assessors and USACHPPM is strongly recommended. The exposure durations associated with these guidelines are 10 minutes, 1 hour, 8 hours, and 24 hours. These durations are presumed to reflect the duration of various military missions and the environmental fate of CWAs (in the air pathways). Estimates of the 24-hour exposure guidelines are based on one-third of the 8-hour level (by using Haber's law). However, for end points that are primarily minimal local effects (e.g., reversible miosis, eye pain), Haber's rule generally is thought not to apply because it overestimates the importance of the duration component. For these local effects, toxicity is determined primarily by the exposure concentration, with the exposure duration ("time") making a smaller contribution. The degree, if any, to which toxicity depends on the duration is not known, and the approach for extrapolating across significantly different exposure durations is not well characterized.

Finally, mechanistic considerations would be informative regarding the conditions and dose metrics for which Haber's law is and is not applicable. A number of mechanistic reasons can be hypothesized as to why the Ct relationships for OP acetylcholinesterase (AChE) inhibitors may not follow Haber's law. First, the rate of AChE inhibition may be an important determinant in the toxicologic response—that is, rapid AChE inhibition may not be tolerated as well as more protracted inhibition. Adaptive processes—for example, receptor desensitization or receptor internalization/down regulation—may occur more effectively with slower AChE inactivation than with rapid inhibition. As G agents and VX are all rapid, direct-acting AChE inhibitors, this may influence the Ct relationship. Finally, synapses are thought to have "spare" AChE molecules—that is, some degree of AChE inhibition can be tolerated at the synapse before cholinergic neurotransmission is functionally affected. This could influence the concentration-duration-response relationship if the rate of synthesis of spare molecules is slower than the rate of inactivation. Other factors that could contribute to the shape of the concentration-duration-response relationship include the potential for regenerative repair.

QUESTION 6. HOW SHOULD EXPOSURE-DURATION RISK BE COMMUNICATED FOR CWAS?

The risk to human health is a function of the concentration and the exposure duration. The best strategy for communicating this risk has yet to be determined. Field commanders ultimately must be able to use the information generated by the DOD Research Plan. Those commanders will only rarely find that traditional formats for presenting toxicologic data meet their needs. Accordingly, one of the challenges of the DOD research program will be to find ways to express or relate relevant data in adequate and useful ways that consider both the need to adequately protect forces from the harmful effects of low levels of CWAs and the offsetting costs of impairing the capacity of forces to accomplish their battle objectives consequent to the use of protective gear or MOPP. DOD risk assessors should confer with operations personnel to determine the nature and form of information that would be most useful. For example, a chart with exposure duration along one axis and exposure concentration along the other axis would allow a readily understood pictorial presentation of the exposure conditions (duration and concentration) associated with various probabilities of adverse effects (e.g., work impairment due to miosis). The committee recommends consultation with risk communication specialists to further assist with this task. Some general remarks related to this topic are given below. At a simple level, a tabular display may suffice. For example, a hypothetical table of the probabilities of adverse response associated with different exposure durations at different CWA concentrations (arbitrary units) is presented (Table 4-1). A figure analogous to this table might also be constructed (see figures at the end of this chapter for different ways to represent these types of data).

Depending on how short, medium, and long exposure durations are defined, it is necessary to describe the range of exposure durations in

TABLE 4-1 Probability of Adverse Response Associated with Various CWA Concentrations and Durations (Hypothetical)

	CWA Concentration				
Exposure Duration	0	10	20	30	40
Short	0	0.40	0.60	0.90	0.90
Medium	0	0.60	0.80	0.95	0.95
Long	0	0.80	0.90	1.00	1.00

which CWA concentration limits are valid as well as a method for adjusting estimates of exposure limits when those ranges are exceeded.

QUESTION 7. WHAT ARE APPROPRIATE APPROACHES FOR EVALUATING THE RISK OF EXPOSURE TO A CWA MIXTURE?

In general, DOD considers the most likely CWA exposures to be those that involve single agents. However, multiple CWAs might be simultaneously deployed. In the development of AEGLs for nerve agents, the National Research Council (NRC) (2003) suggested that the effects of concurrent exposures to commercial insecticides and G agents would be dominated by the more potent nerve agents. This conclusion is based on large differences in lethal concentration in 50% of exposed individuals (LC_{50}) values among the G agents, agent VX, and commercial insecticides. The NRC (2003) further pointed out that when administered together in laboratory animals, the toxicity of two G-series agents was "basically additive" (Clement 1994; Luo and Liang 1997), which is expected for compounds sharing anticholinesterase properties. No potentiation has been observed in toxicologic tests of nerve agent mixtures.

Unless detection methods are available to discriminate among agents in the field, it is reasonable to focus on the most potent chemical for setting exposure guidelines for mixtures relevant to battlefield conditions. The committee assumes that if a mixture were present, it would be composed of a relatively small number of CWAs. Existing data characterizing the toxicity of nerve agent mixtures indicate that the toxicity of the mixture would approximate that of the most potent component. Thus, it is reasonable to evaluate the risk for a nerve agent mixture assuming that the entire mixture was composed of its most potent constituent.

When detection methods to discriminate among different agents are available, standard methods for mixture risk assessment can be used. For chemicals that act via similar modes of action, the default approach for assessing the risk from combined exposure is dose addition. For nerve agents, experimental data collected from administration of a G-agent mixture confirms the validity of this approach (reviewed by NRC 2003). This approach adds the scaled doses of the different components of the mixture, accounting for relative toxicity. Dose-additive approaches include the hazard index, toxicity equivalence factor, and relative potency factors.

Existing and well-conducted experimental studies indicate that the toxicity of G-series nerve agents is additive. Relative-potency analyses documented in the NRC (2003) report support this determination for nerve agent VX as well. Consideration of other available data could define the degree of toxicity independence between and among CWAs possessing different mechanisms (e.g., alkylating agents vs. anticholinesterase agents); this area would benefit from further exploration. The results of such exploration would inform the need and direction for any further model considerations. Such an approach would support the application of ORM, which considers the likelihood of adverse effects associated with various exposures.

QUESTION 8. WHAT INFORMATION IS NEEDED TO BEST SUPPORT EXTRAPOLATION FROM ANIMAL STUDIES TO HUMAN RESPONSES? HOW AND WHEN MIGHT A PBPK MODEL BE USEFUL FOR AIDING IN SUCH EXTRAPOLATIONS?

Determining the probability of a toxic response from human data is straightforward if the population for which data exist is comparable to the population of interest and if data are available for the exposure duration of interest. (See Question 5.) Methods for extrapolating threshold levels from animal data are well defined (albeit there are uncertainties inherent in the approach), but extrapolation of response probabilities is more complex. Essentially, it is often assumed that humans have the same probability of response as animals when both receive an equivalent dose, although this approach does not address interspecies differences in sensitivity to a given dose (due to differences in tissue susceptibility, repair, defense, and other factors). The definition of this "equivalent dose" is the challenge of species extrapolation. Hence, establishment of species equivalency in response, or lack thereof, is critical. Biologically based dose-response modeling makes it possible to predict probabilities in humans, but it is very labor intensive. PBPK modeling can aid in extrapolating exposures to humans. These models incorporate information on the physiology and kinetics of metabolism of the chemical of interest in humans and the experimental animal species and so allow for determination of the human exposure that results in the same tissue dose as received by the experimental animals. As an important aside, however, a PBPK model may not be required for extrapolating a toxic response from a direct contact (e.g., miosis may be an example of this if direct contact of the CWA with the eye causes this response).

A key issue when a PBPK model is used is identification of the appropriate measure of tissue dose, the "dose metric." The correct dose metric will describe the dose response for the chemical under multiple sets of exposure conditions. Common dose metrics include, but are not limited to, the area under the curve of the chemical or its metabolite in the plasma or target tissue, peak concentrations, and amount metabolized. The use of any animal model in risk assessment would require considering whether the model has been adequately validated in the experimental animal species, whether it has been validated for humans, and whether the mode of toxicity for CWAs in the modeled species is relevant to other mammals, notably humans. Maxwell et al. (1988) developed a physiologically based pharmacodynamic (PBPD) model for GD in the rat, while Gearhart et al. (1990, 1994) developed a similar combined PBPK/PBPD model for the OP diisopropyl fluorophosphate (DFP), a model OP with some resemblance to nerve agents. This model allows one to estimate not only tissue levels of DFP but also the toxic response in terms of AChE levels in the brain or plasma. More recent work is exploring the use of fluoride ion regeneration as a dose metric that is more sensitive and less variable than changes in AChE at low exposure levels. Further work in extending this sort of model to other CWAs will be useful in improving the interspecies extrapolation for CWAs as well as in addressing exposure duration extrapolation issues. For example, Yu et al. (2004) have reported on the preliminary development of a PBPK/PBPD model for GB in miniature swine to aid in interpretation of results from toxicity studies with swine. The model includes an ocular compartment, including ocular absorption of GB and binding to AChE, to aid in addressing miosis. Incorporating data on human variability may make it possible to predict the proportion of exposed humans expected to be adversely affected.

The toxicokinetics of GD is complicated by the high in vivo reactivity and distinct differences in metabolic properties of the toxic C(+/−)P(−)-stereoisomers and the less toxic C(+/−)P(+)-stereoisomers. For VX, the overall stereospecificity for sequestration of (+)- versus (−)-VX is much less pronounced than for the stereoisomers of G agents, based on the observed marked in vivo persistence of VX. The first PBPK model that took into consideration the chirality of GD was by Langenberg et al. (1997; cited by Benschop and De Jong 2001). In this model, binding in blood becomes less dominant in overall toxicokinetics when the dose of C(+/−)P(−)-GD exceeds the capacity of the binding sites in blood. This highlights the need for further refinement by discriminating among the four stereoisomers of GD.

Benschop and De Jong (2001) and the NRC (2003) reviewed the data on interspecies differences. In view of the large differences in susceptibilities to G agents between rats and guinea pigs and the relative role of plasma carboxylesterases (i.e., these esterases are found in relatively high concentrations in rat but not in guinea pig plasma), the guinea pig is often considered a better model for these toxicants. Carboxylesterases do not play a significant role in detoxification of VX, however, leading the committee to conclude that rats are not a less-appropriate model with this nerve agent. Thus, the most appropriate animal model to use based on these species differences in agent detoxification can vary with the particular agent in question.

Although biologically based dose-response modeling allows one to predict response in humans in response to a specified external exposure, PBPK modeling calculates only tissue dose. Additional extrapolation is needed to consider toxicodynamics (pharmacodynamics) or interspecies differences in tissue response. Several of the existing PBPK models for CWAs include a pharmacodynamic component (Maxwell et al. 1988; Gearhart et al. 1990; Yu et al. 2004), addressing, for example, how AChE levels change with tissue dose. This addresses a major component of the pharmacodynamic differences, although it does not address any interspecies differences in operationally relevant performance decrements for a given AChE level. U.S. and international agencies typically use a factor of 3 or 3.2 (half the default factor of 10 on a log scale) to address interspecies differences in toxicodynamics when data from a PBPK model are used to replace the toxicokinetic portion of the interspecies uncertainty factor (IPCS 2001). Toxicologic judgment could be used to reduce this factor for toxicodynamic differences to a factor of 1 if the critical effect is in a species that is judged to react very similarly to humans, such as many end points in nonhuman primates. Similarly, the use of validated PBPK/PBPD models, when available for the species and end point of interest, would address most, if not all, of the interspecies differences in toxicodynamics, markedly reducing or eliminating the need for uncertainty factors to address this area of uncertainty. NRC (2003) notes that the miotic response of the mammalian eye appears to be quantitatively very similar across species, including humans, indicating that no toxicodynamic uncertainty factor is needed for this end point.

QUESTION 9. WHAT IS THE IMPACT OF EXPERIMENTAL DESIGNS ON RISK ASSESSMENTS AND SUBSEQUENT EFFORTS TO ESTABLISH EXPOSURE GUIDELINES FOR MILITARY PERSONNEL PROTECTION?

The lack of experimental data for longer exposure durations (e.g., 8 hours, 24 hours) and at lower concentrations is an acknowledged limitation in efforts to establish military exposure guidelines for exposure durations greater than 8 hours. As a consequence, the protective assumption of response linearity from 8 to 24 hours was applied by USACHPPM to develop 24-hour military exposure guidelines for nerve and mustard agents from the 8-hour AEGL (e.g., assuming Haber's law for exposure duration greater than 8 hours; it should be noted that Haber's law is a special case of the ten Berge expression when $n = 1$). The 8-hour AEGLs were time-scaled using the techniques described by NRC (2001) from human and laboratory animal experimental exposure durations of less than 8 hours, assuming the ten Berge et al. (1986) expression of dose-response and chemical-specific determination of n. For example, the 8-hour AEGL-1, AEGL-2, and AEGL-3 were extrapolated from 4-hour, 30-minute, and 6-hour experimental data, respectively. The agent GB data set is robust (NRC 2003) and supports the assumptions used. As a general rule, the greater the extrapolation from the original data, the greater the resulting uncertainty. Grotte and Yang (2001) had speculated that inhalation concentration for the G agents probably could be extrapolated from 2 minutes through 60 minutes with reasonable confidence. Without access to the more recent experimental data of Mioduszewski et al. (2000, 2001, 2002a, 2002b) and of van Helden et al. (2001, 2002, 2003), Grotte and Yang (2001) further speculated that the accuracy of extrapolating below 2 minutes and beyond 60 minutes is unknown. This assumption has been superseded by current assessments with more recent nerve agent experimental data (NRC 2003). Model extrapolation beyond the tested concentration-time scenarios can be done with reasonable precautions. Experiments that include durations of exposure and concentrations that are operationally relevant are preferred. Conducting studies at operationally relevant levels helps alleviate extrapolation in the risk characterization of low-level exposures to CWAs.

To operationalize the ORM concept, the risk assessor will require information on response probabilities from the experimental study. For each critical response associated with CWA exposure, the research should study the complete dose-duration-response profile and not simply some single end point, such as ECt_{50} (the concentration and time that causes an effect in 50% of subjects) and EC_{01} (concentration effective in 1% of subjects) for a specific effect for a specified exposure duration. Access to recently generated experimental data sets and contact with investigators should be pursued to develop other response probabilities (e.g., 15%, 30%, 40%) that would be needed in operational risk assessment. Experimental design consideration of adequate size for each exposure-response group would help minimize uncertainties in subsequent extrapolations.

QUESTION 10. HOW DOES ONE ACCOUNT FOR VARIABILITY IN HUMAN RESPONSE TO CWAS, AND DO CWA-SUSCEPTIBLE SUBPOPULATIONS OF HUMANS EXIST?

Factors contributing to human variability depend on the end point and the mechanistic basis for the end point. Miosis that can occur from local exposure in the absence of systemic cholinesterase inhibition is a local effect. For such effects, variability would be determined by the chemistry and physiology of the eye, including metabolic enzymes in the eye that can deactivate nerve agents, and the variability in response in the neurons controlling miosis.

For systemic effects of nerve agents, variability in human response can be due to differences in the kinetics of agent uptake and metabolism or to differences in susceptibility to a given tissue dose. Thus, it can be predicted that variability in response to systemic exposure would be higher than that involving local effects. Both plasma cholinesterase and red blood cell AChE activities generally are lower in women than in men by about 10%. This means that a given percent decrease in cholinesterase generally results in a lower absolute activity for women, so that there is a smaller "buffer" between normal cholinesterase levels and levels that can result in toxicity. It has also been shown that a small subpopulation of men and women have genetically determined variants in their plasma cholinesterase resulting in very low activity levels. Individuals with these genetic variants may be unusually susceptible to some anticholinesterase compounds (Young et al. 1999). As reviewed by NRC (2003), the frequency of the homozygous pheno-type is estimated at 0.025% (Hayes 1982), and plasma cholinesterase activity is less than 25% of normal in these individuals (Bonderman and Bonderman 1971).

Variability in nerve agent kinetics is due to genetic factors, such as polymorphisms in the paraoxonase gene (PON1), as well as variability in expression of the relevant enzymes (Furlong et al. 2002) and variability in breathing rate, which affects the amount of agent taken up into the body. As discussed above, variability in breathing rate may be particularly important for military personnel. Young animals have lower levels of paraoxonase, as well as carboxylesterase, but this age-related difference in susceptibility is not of concern in the context of military operational performance.

Because many of the sources of variability would also apply to military populations, the committee considers this group likely to be as variable as the general population. Further characterization of the implications of human variability in response to CWAs under military deployment conditions would be useful for ORM.

QUESTION 11. WHAT STATISTICAL MODELS ARE APPROPRIATE FOR CONCENTRATION-TIME DATA?

The modeling of dichotomous responses in dose-duration studies is discussed below. The committee's focus is on the evaluation of data from an experimental study of a single agent. In particular, the structure imposed by Haber's law and possible generalizations are described. As mentioned above, Haber's Law states that cumulative exposure determines the toxic response (e.g., an exposure at 100 parts per million [ppm] for 6 hours leads to the same toxic response as 600 ppm for 1 hour). To illustrate the impact of Haber's law on a statistical model, we focus on a dichotomous response (e.g., the presence of significant impairment). For a dichotomous response, the probability of a material impairment is modeled as a function of dose (or concentration) and duration (or time). Denoting the concentration by C and duration by t, the probability of an adverse response given exposure to concentration C of an agent for t units of time is $\pi(C, t) = \text{function}(C, t)$. A generalized linear model (McCullagh and Nelder 1989) with a logit link or a probit link and a binomial response distribution is commonly used in this situation. In this case, under Haber's law, the model can be written as follows:

$$\text{logit}[\pi(C, t)] = \alpha + \beta\,(C \times t), \qquad (1)$$

where $\text{logit}(x) = \log[x/(1 - x)]$. A contour plot of this pattern is shown in Figure 4-1. The "X" points in this figure provide one option of a factorial experimental design where combinations of concentrations and durations are tested. Table 4-2 provides an example of a better experimental design that more specifically targets the experimental data points to provide the most information on the dose-response.

As an aside, this model provides a prediction of responses in an animal study. In general, extrapolation to humans may require other adjustments, although, as observed previously, this may not be an issue for miosis caused by direct contact with CWAs.

The contour plot in Figure 4-1 corresponds to a projection of the three-dimensional surface (Figure 4-2). The region in the top right corner of the plot (region in the Ct plan located above the line labeled "0.95") corresponds to concentration-duration combinations resulting in >95% of adverse response.

Note that the models shown do not include a mathematical threshold, even though such thresholds are assumed biologically for the primary effects of concern for CWAs. The absence of a mathematical threshold would not have a substantial impact on the model estimates in

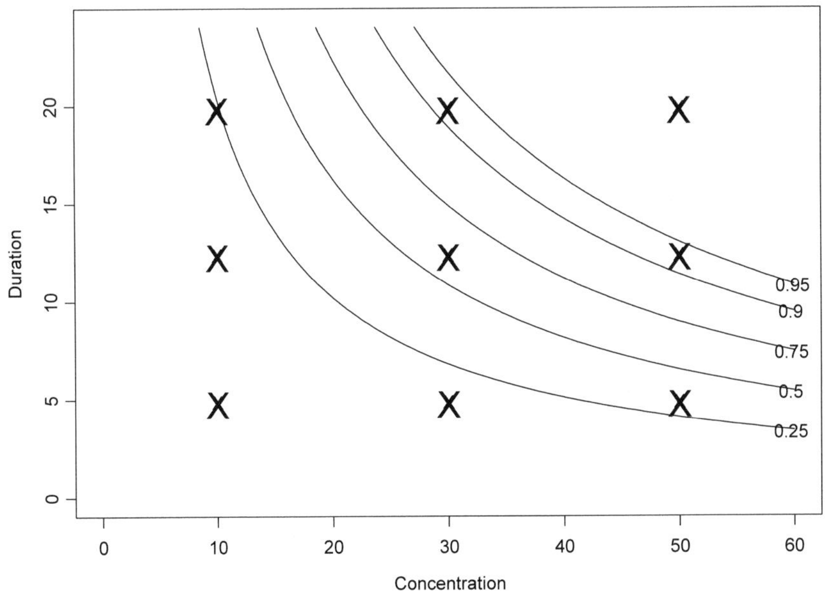

FIGURE 4-1 Contour plot of hypothetical data assuming that Haber's law applies (using parameter values $\alpha = -2.944439$, $\beta = 0.009036316$ in equation 1). The probability of adverse response is displayed as a function of concentration and duration.

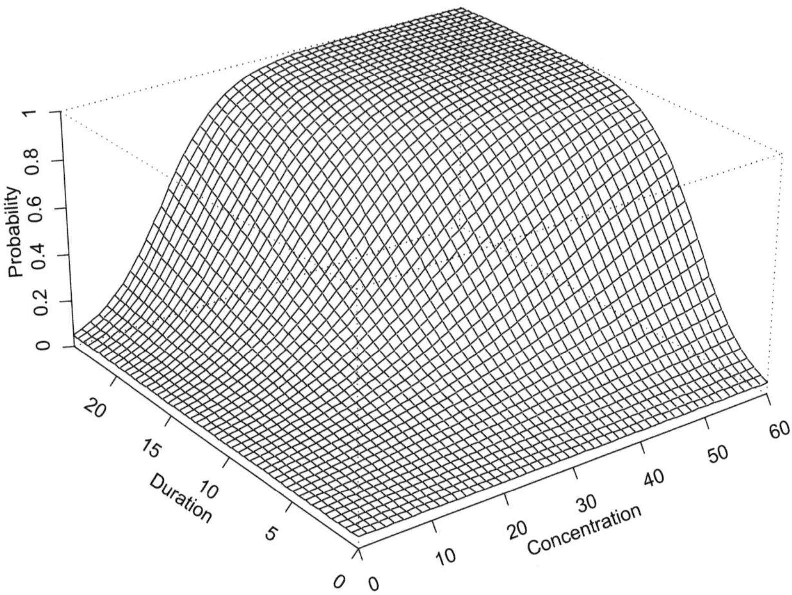

FIGURE 4-2 Perspective plot of hypothetical data assuming Haber's law applies (using parameter values α = −2.944439, ß = 0.009036316 in equation 1).

TABLE 4-2 Example of Experimental Design[a] That Might Be Considered an Alternative to a Factorial Design

Duration (t)	C = 1	C = 3	C = 9
16	X		
8	X	X	
4	X	X	X
2		X	X
1		X	

[a]Arbitrary units are for illustration. X indicates a tested duration-concentration combinations, and Ct ranges from 4 to 36.

the range of the data, but these models should not be used to extrapolate response predictions to conditions well below the range of the data.

Departures from this model often can be incorporated by adding an additional time term to the model. For example, this is illustrated with the following logistic regression model:

$$\text{logit}[\pi(C, t)] = \alpha + \beta (C \times t) + \gamma t, \quad (2)$$

where α, β, and γ are variable parameters used to fit data.

An illustration of a hypothetical contour plot when Haber's law does not apply ($\gamma \neq 0$) is given below. Note that this model can be expressed as follows:

$$\text{logit}[\pi(C, t)] = \alpha + (\beta C + \gamma) t \quad (3)$$

and highlights one possible modification of the pattern imposed by Haber's law. In this case, if $\gamma > 0$, then toxicity greater than predicted by Haber's law is observed. This is illustrated in Figures 4-3 and 4-4. As an aside, this model assumes that the probability of an adverse response increases with time even in unexposed individuals ($C = 0$). Model 3 implies $\text{logit}[\pi(C = 0,t)] = \alpha + \gamma t$. If the probability of adverse response is assumed constant in the absence of dose ($C = 0$) and Haber's law does

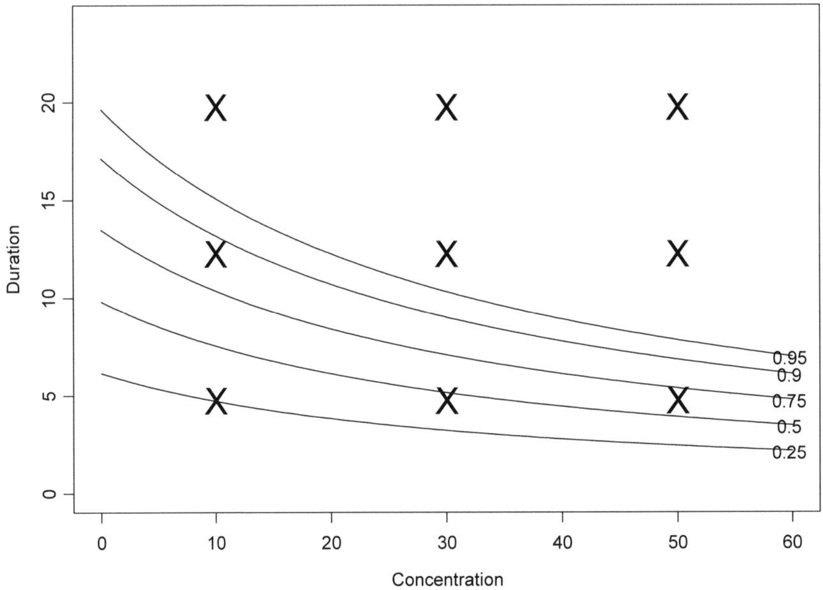

FIGURE 4-3 Contour plot of hypothetical data assuming Haber's law does not apply (using parameter values $\alpha = -2.944439$, $\beta = 0.009036316$, $\gamma = 0.3$ in equation 2). The probability of adverse response is displayed as a function of concentration and duration.

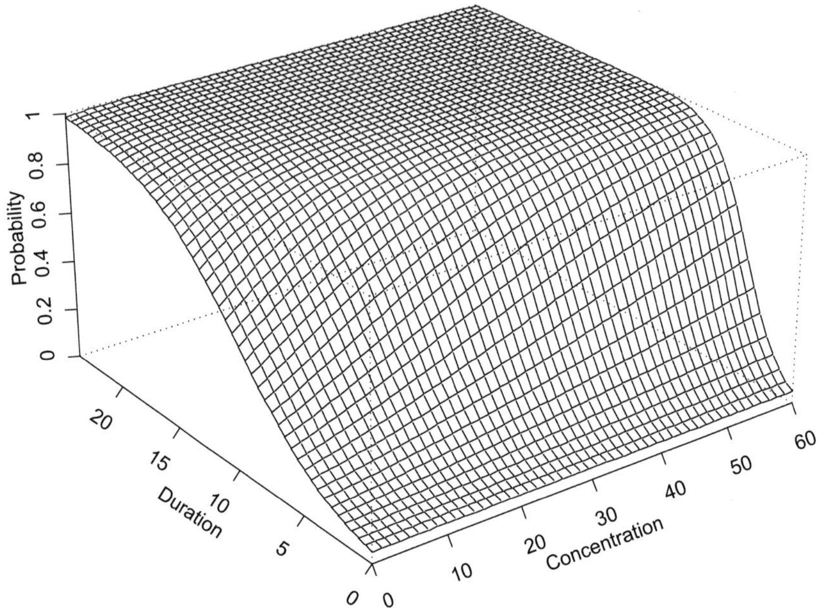

FIGURE 4-4 Perspective plot of hypothetical data assuming Haber's law does not apply (using parameter values $\alpha = -2.944439$, $\beta = 0.009036316$, $\gamma = 0.3$ in equation 2).

not apply, then an alternative model might be written as $\text{logit}[\pi(C, t)] = \alpha + \beta C (1 + \gamma) t$ or with ten Berge's modification $\text{logit}[\pi(Cit)] = \alpha + \beta C^n t$.

The "X" points in the figure provide the same hypothetical sample design points as shown in Figure 4-1, assuming that $10 \leq$ concentrations ≤ 60 and $5 \leq$ durations ≤ 20 (units arbitrary) are of interest. However, unlike the results in Figure 4-1, this sampling design would result in most of the data being collected at concentration-duration combinations that result in ≥95% response. In this case, one should conduct range-finding studies to better define the concentration-duration range of interest or test lower concentrations for the exposure durations of interest.

This contour was associated with the three-dimensional surface shown in Figure 4-4. This full model reduces to the model suggested by Haber's law when $\gamma = 0$. What do these figures illustrate about Haber's law being violated? If one compares the contour plots (Figures 4-1 and 4-3), one sees that for constant values of α and β, much lower exposures for the same duration lead to higher probabilities of adverse response when Haber's law does not apply. For example, a concentration = 20 × duration = 10 (units arbitrary) exposure is associated with an estimated probability of adverse response of <0.25 when Haber's law is assumed; however, this same concentration-duration pattern is associated with >0.75 probability of adverse response when Haber's law is not operative. The perspective plots (Figures 4-2 and 4-4) confirm the observation that higher probabilities of adverse response are predicted for each concentration-time exposure combination. The implication of this observation is that the naive and incorrect application of Haber's law to the setting of CWA exposure guidelines may lead to erroneous and, in this example, unprotective risk estimates.

While many investigators use the available experimental data to calculate the coefficients α and ß, others would (erroneously) apply Haber's law directly to extrapolate from concentration-response data at one time point to other time points. The implications for prediction of response at other concentration-duration combinations can be considered by using ten Berge's generalization of this relationship, which states that Cn × t is a constant. Values of n > 1 tend to flatten the concentration-time contour plot, making the plot more like Figure 4-3 and less like Figure 4-1. Again, focusing on the contour plot, this means that as one moves to shorter durations from the available data, erroneously applying Haber's law when n > 1 would tend to overestimate the concentration one can be exposed to for a given response contour and thus would not be a health-protective approach. Conversely, erroneously applying Haber's law in extrapolating to longer exposure conditions when n > 1 would tend to underestimate the concentration one can be exposed to for a given response contour and thus would be overprotective.

REGRESSION METHODS FOR CONTINUOUS OR ORDERED RESPONSES

The analyses presented above assumed that adverse response was dichotomous (e.g., impaired or not impaired). Pupillary dilation could be measured as a continuous parameter and standard regression techniques (Neter et al. 1996) might be used. In other words, if pupillary constriction is a measure on a continuous scale, then the response can be used in a normal-theory regression model. Equation 1 and extensions would be modified accordingly. For a continuous response, say, Y = continuously measured pupillary constriction,

$$Y = \alpha + \beta\,(Cn \times t) + \text{error}. \qquad (4)$$

Note that the assumed structure for the relationship of the response and dose and duration should be evaluated for any response. Finally, if severity scales are used for the response—that is, the response may be categorized into a set of ordered categories—then ordinal response regression methods should be considered (Agresti 1984).

EXPERIMENTAL RECOMMENDATIONS

For detecting deviations from Haber's law, a minimum of two concentrations multiplied with two durations resulting in four Ct treatments would be required. The committee recommends conducting studies with a minimum of nine Ct treatments (three concentrations multiplied by three durations). In general, the committee believes that if models are to be used to predict risk at a variety of concentrations and durations, then more extensive data are required to capture the nuances of the relationship between response, concentration, and exposure duration. Such relationships are not well established for CWAs.

THE COMMITTEE'S RECOMMENDATIONS

Although miosis typically has been considered the critical effect and most frequently used indicator of toxicity, questions remain about other adverse effects at low-level exposures. DOD should conduct research to identify whether there are more sensitive toxicity end points, other than miosis, from exposures to low levels of CWAs.

If miosis is selected as the critical adverse response, studies must determine what level of pupillary constriction in miosis is operationally relevant. This information is needed for ORM risk-risk comparison, which must consider the range of exposure limits for and probabilities of adverse outcomes from exceedances of those limits for each level of severity and operationally relevant exposure duration.

Although the uses of biomarkers may have limited relevance to immediate ORM decisions, they might be useful for research on postexposure sequelae, refining interspecies extrapolations, and exposure reconstruction. Future research should focus on the practical application of the biomarker of interest—that is, whether samples taken weeks after exposure can be used to identify and quantify exposure.

Military personnel are likely to be exposed to low levels of CWAs while physically active. The use of animal models in which exposures are conducted while the animals are physically restrained may be problematic unless PBPK/PBPD models are used to address the implications of differences in breathing rates. Even then, care needs to be taken to ensure that measured effects are not due to stress or interaction between the agent and the stress. Appropriate consideration of extreme environmental conditions and high activity levels during deployment needs to be taken in future research.

The setting of exposure limits for CWAs usually has involved extrapolation to other concentrations or durations by invoking Haber's law (adverse response is related to the product of agent concentration and duration of exposure) or ten Berge calculation, a generalization of Haber's law that can (1) weight concentration more than time when $n > 1$, (2) weight concentration less than time when $n < 1$, or (3) weight concentration and time equally. Naive applications of Haber's law can lead to erroneous risk estimates. PBPK/PBPD modeling should be used to improve evaluation of concentration-duration-response relationships. Mechanistic considerations would also be informative regarding the conditions and dose metrics for which Haber's law is and is not applicable. Studies that allow for the testing and validation of duration-exposure models are encouraged.

Ask combat doctrine developers to list tactical decision requirements and then ask researchers to identify which requirements would be addressed by proposed research. This should not discourage basic research but should help keep toxicologists focused on the most critical problems demanding ultimate solutions.

Field commanders ultimately must be able to use the information generated by this research program to make decisions. The DOD toxicology establishment should confer with operations personnel to determine the nature and form of information that would be most useful. The committee recommends consultation with risk communication specialists to further assist with this task.

There is a possibility that multiple CWAs could be used simultaneously. Unless methods are available to discriminate among agents in the field, it is reasonable to focus on the most potent agent for setting exposure guidelines for mixtures relevant to battlefield conditions and assume that the entire mixture would be less potent than an equal exposure to the most potent component of the mixture.

PBPK modeling can aid in extrapolating exposures from animals to humans. Recent work exploring the use of fluoride ion regeneration as a dose metric that is more sensitive and less variable than changes in AChE may be useful in improving the interspecies extrapolation for CWAs as well as addressing the exposure duration extrapolation issue and should be further explored. It should be noted that a PBPK model may not be required for extrapolating a direct contact toxic response (e.g., miosis may be an example of this if direct-contact of the CWA with the eye leads to this response).

To make the ORM concept operational, the risk assessor will require robust information on response probabilities from the experimental study. For CWAs, for each critical response (e.g., miosis), access to recently generated experimental data sets and contact with investigators should be pursued to develop other response probabilities that would be needed in operational risk assessment beyond single values, such as ECt_{50} or EC_{01} (e.g., 15%, 30%, 40%).

The toxicity end point and the mechanistic basis for the end point are factors contributing to human variability. For miosis alone, variability would be determined by the chemistry and physiology of the eye. Ambient light conditions can also contribute to the variability in miosis. For systemic effects of CWAs, variability in human response can be due to differences in the kinetics of CWA uptake and metabolism or differences in sensitivity to a given tissue dose of CWA. Further characterization of the implications of human variability in response to CWAs under military conditions would be useful for ORM.

The methodology developed for deriving AEGL values for CWAs is pertinent. The generalization of Haber's law to effects related to $C^n t$ (ten Berge method) is a valuable contribution. The committee recommends that DOD utilize information and techniques developed for deriving AEGLs.

References

Agresti, A. 1984. Analysis of Ordinal Categorical Data. New York: John Wiley.

Anthony, J.S., M.V. Haley, J.H. Manthei, R.A. Way, D.C. Burnett, B.P. Gaviola, D.R. Sommerville, R.B. Crosier, R.J. Mioduszewski, E.M. Jakubowski, J.L. Montgomery, and S.A. Thomson. 2002. Inhalation Toxicity of GF Vapor in Rats as a Function of Exposure Concentration and Duration and its Potency Comparison to GB. Late-breaking poster presented at 41st Annual Meeting of the Society of Toxicology, March 21, 2002, Nashville, TN.

Atherley, G. 1985. A critical review of time-weighted average as an index of exposure and dose, and of its key elements. Am. Ind. Hyg. Assoc. J. 46(9):481-487.

Baker, D.J., and E.M. Sedgwick. 1996. Single fibre electromyographic changes in man after organophosphate exposure. Hum. Exp. Toxicol. 15(5):369-375.

Benschop, H.R., and L.P.A. DeJong. 2001. Toxicokinetics of nerve agents. Pp. 25-81 in Chemical Warfare Agents: Toxicity at Low Levels, S.M. Somani, and J.A. Romano, eds. Boca Raton, FL: CRC Press.

Black, R.M., R.J. Clarke, J.M. Harrison, and R.W. Read. 1997. Biological fate of sulfur mustard: Identification of valine and histidine adducts in haemoglobin from casualties of sulfur mustard poisoning. Xenobiotica 27(5):499-512.

Blick, D.W., M.R. Murphy, G.C. Brown, M.G. Yochmowitz, J.W. Fanton, and S.L. Hartgraves. 1994. Acute behavioral toxicity of pyridostigmine or soman in primates. Toxicol. Appl. Pharmacol. 126(2):311-318.

Bomser, J.A., and J.E. Cassida. 2001. Diethylphosphorylation of rat cardiac M2 muscarinic receptor by chlorpyrifos oxon in vitro. Toxicol. Lett. 119(1):21-26.

Bonderman, R.P., and D.P. Bonderman. 1971. Atypical and inhibited human serum pseudocholinesterase: A titrimetric method for differentiation. Arch. Environ. Health 22(5):578-581.

Brown, M.A., and K.A. Brix. 1998. Review of health consequences from high-, intermediate- and low-level exposure to organophosphorus nerve agents. J. Appl. Toxicol. 18(6):393-408.

Bucci, T.J., R.M. Parker, and P.A. Gosnell. 1992a. Toxicity Studies on Agents GB and GD (Phase II): Delayed Neuropathy Study of Soman in SPF White Leghorn Chickens. AD-A257183. Prepared for the U.S. Army Biomedical Research and Development Laboratory, Fort Detrick, MD, by the National Center for Toxicological Research, Jefferson, AR.

Bucci, T.J., R.M. Parker, J.A. Crowell, J.D. Thurman, and P.A. Gosnell. 1992b. Toxicity Studies on Agent GA (Phase II): 90 Day Subchronic Study of GA (Tabun) in CD Rats. Final Report. AD-A258020. Prepared for the U.S. Army Biomedical Research and Development Laboratory, Fort Detrick, MD, by the National Center for Toxicological Research, Jefferson, AR.

Burchfiel, J.L., F.H. Duffy, and M. van Sim. 1976. Persistent effects of sarin and dieldrin upon the primate electroencephalogram. Toxicol. Appl. Pharmacol. 35(2):365-379.

Castro, C.A., V.C. Gresham, A.V. Finger, D.M. Maxwell, R.P. Solana, and D.E. Lenz. 1994. Behavioral decrements persist in rhesus monkeys trained on a serial probe recognition task despite protection against soman lethality by butyrylcholinesterase. Neurotoxicol. Teratol. 16(2):145-148.

CBIAC (Chemical and Biological Defense Information Analysis Center). 1995. Worldwide Chemical Detection Equipment Handbook. HB-95-02. CBIAC Gunpowder Branch Aberdeen Proving Ground, MD.

CBIAC (Chemical and Biological Defense Information Analysis Center). 1998a. The Emergency Responder's Ability to Detect Chemical Agents. CR-98-06. CBIAC Gunpowder Branch Aberdeen Proving Ground, MD.

CBIAC (Chemical and Biological Defense Information Analysis Center). 1998b. Assessment of Chemical Detection Equipment for Hazmat Responders. SOAR-98-05. CBIAC Gunpowder Branch Aberdeen Proving Ground, MD.

Chemical Casualty Care Office. 1995. Medical Management of Chemical Casualties Handbook, 2nd Ed. Chemical Casualty Care Office, Medical Research Institute of Chemical Defense, Aberdeen Proving Ground, MD. [Online]. Available: http://www.fas.org/nuke/guide/usa/doctrine/army/mmcch/ [accessed June 16, 2004].

Chilcott, R.P., C.H. Dalton, I. Hill, C.M. Davidson, K.L. Blohm, and M.G. Hamilton. 2003. Clinical manifestations of VX poisoning following percutaneous exposure in the domestic white pig. Hum. Exp. Toxicol. 22(5):255-261.

Clement, J.G. 1994. Toxicity of the combined nerve agents GB/GF in mice: Efficacy of atropine and various oximes as antidotes. Arch. Toxicol. 68(1):64-66.

Diggle, P., K.Y. Liang, and S.L. Zeger. 1994. Analysis of Longitudinal Data. Oxford, England: Clarendon Press.

DOD (U.S. Department of Defense). 2003. Low-Level Chemical Warfare Agents (CWAs) Exposure Research Master Plan. June 2003.

Duffy, F.H., J.L. Burchfiel, P.H. Bartels, M. Gaon, and V.M. Sim. 1979. Long-term effects of an organophosphate upon the human electroencephalogram. Toxicol. Appl. Pharmacol. 47(1):161-176.

EPA (U.S. Environmental Protection Agency). 2000. Supplementary Guidance for Conducting Health Risk Assessment of Chemical Mixtures. EPA/630/R-00/002. Risk Assessment Forum, U.S. Environmental Protection Agency, Washington, DC. August 2000. [Online]. Available: http://www.epa.gov/ncea/raf/pdfs/chem_mix/chem_mix_08_2001.pdf [accessed June 21, 2004].

EPA (U.S. Environmental protection Agency). 2003. Catreg Software for Categorical Regression Analysis. National Center For Environmental Assessment, Office of Research and Development, U.S. Environmental Protection Agency, Research Triangle Park Office, Research Triangle Park, NC [online]. Available: http://permanent.access.gpo.gov/lps35390/cfpub.epa.gov/ncea/cfm/recordisplay.cfm-deid=18162.htm [accessed Sept. 14, 2005].

EPA (U.S. Environmental Protection Agency). 2004. Glossary of IRIS Terms. Integrated Risk Information System, U.S. Environmental Protection Agency. [Online]. Available: http://www.epa.gov/iris/gloss8.htm#c [accessed May 2004].

Ferraro, F.R., and M. Gabriel. 2003. Preservation of implicit memory in adult children of alcoholics. J. Psychol. 137(4):373-380.

Fidder, A., A.G. Hulst, D. Noort, R. de Ruiter, M.J. van der Schans, H.P. Benschop, and J.P. Langenberg. 2002. Retrospective detection of exposure to organophosphorus anti-cholinesterases: Mass spectrometric analysis of phosphylated human butyrylcholinesterase. Chem. Res. Toxicol. 15(4):582-590.

Furlong, C.E., W.F. Li, D.M. Shih, A.J. Lusis, R.J. Richter, and L.G. Costa. 2002. Genetic factors in susceptibility: Serum PON1 variation between individuals and species. Hum. Ecol. Risk Asses. 8(1):31-44.

GAO (U.S. Government Accounting Office). 1998. Chemical Weapons: DOD Does Not Have a Strategy to Address Low-Level Exposures: Report to Congressional Requestors. GAO/NSIAD-98-228. Washington, DC: U.S. Government Accounting Office. September 1998.

Gaylor, D.W., and W. Slikker, Jr. 1990. Risk assessment for neurotoxic effects. Neurotoxicology 11(2):211-218.

Gearhart, J.M., G.W. Jepson, H.J. Clewell, III, M.E. Andersen, and R.B. Conolly. 1990. Physiologically based pharmacokinetic and pharmacodynamic model for the inhibition of acetylcholinesterase by diisopropylfluorophosphate. Toxicol. Appl. Pharmacol. 106(2):295-310.

Gearhart, J.M., G.W. Jepson, H.J. Clewell, M.E. Andersen, and R.B. Conolly. 1994. Physiologically based pharmacokinetic model for the inhibition of acetylcholinesterase by organophosphate esters. Environ. Health Perspect. 102(Suppl. 11):51-60.

Gentry, P.R., C.E. Hack, L. Haber, A. Maier, and H.J. Clewell III. 2002. An approach for the quantitative consideration of genetic polymorphism data in chemical risk assessment: Examples with warfarin and parathion. Toxicol. Sci. 70(1):120-139.

Goldman, M., A.K. Klein, T.G. Kawakami, and L.S. Rosenblatt. 1987. Toxicity Studies on Agents GB (Sarin-Types I and II) and GD (Soman), Final Report. AD-A187841. Prepared for U.S. Army Medical Research and Development Command, Fort Detrick, MD, by Laboratory for Energy-Related Health Research, University of California, Davis.

Gray, P.J., and R.M. Dawson. 1987. Kinetic constants for the inhibition of eel and rabbit brain acetylcholinesterase by some organophophates and carbamates of military significance. Toxicol. Appl. Pharmacol. 91(1):140-144.

Grotte, J.H., and I.L. Yang. 2001. Report of the Workshop on Chemical Agent Toxicity for Acute Effects, May 11-12, 1998, Institute for Defense Analysis. IDA Document D-2176. Institute for Defense Analyses. Alexandria, VA [Online]. Available: https://usachppm.apgea.army.mil/chemicalagent/caw/IDAreport(2001).pdf [accessed June 16, 2004].

Gurevitch, J., and L.V. Hedges. 1993. Meta-analysis: Combining the results of independent experiments. Pp. 378-401 in Design and Analysis of Ecological Experiments, S.M. Schenier, and J. Gurevitch, eds. London: Chapman and Hall.

Guth, D.J., R.J. Carroll, D.G. Simpson, and H. Zhou. 1997. Categorical regression analysis of acute exposure to tetrachloroethylene. Risk Anal. 17(3):321-332.

Haber, L., J.A. Strickland, and D.J. Guth. 2001. Categorical regression analysis of toxicity data. Comments Toxicol. 7(5/6):437-452.

Hartgraves, S.L., and M.R. Murphy. 1992. Behavioral effects of low-dose nerve agents. Pp. 125-154 in Chemical Warfare Agents, S.M. Somani, ed. San Diego, CA: Academic Press.

Hayes, W.J. 1982. Pesticides Studied in Man. Baltimore, MD: Lippincott William and Wilkins.

Hedges, L.V., and I. Olkin. 1985. Statistical Methods for Meta Analysis. Orlando, FL: Academic Press.

Henderson, R.F., E.B. Barr, W.B. Blackwell, C.R. Clark, C.A. Conn, R. Kalra, T.H. March, M.L. Sopori, Y. Tesfaigzi, M.G. Ménache, and D.C. Mash. 2002. Response of rats to low levels of sarin. Toxicol. Appl. Pharmacol. 184(2):67-76.

Hertzberg, R.C. 1989. Fitting a model to categorical response data with application to species extrapolation of toxicity. Health Phys. 57(Suppl. 1):405-409.

Hertzberg, R.C., and M. Miller. 1985. A statistical model for species extrapolation using categorical response data. Toxicol. Ind. Health 1(4):43-57.

Himuro, K., S. Murayama, K. Nishiyama, T. Shinoe, H. Iwase, M. Nagao, T. Takatori, and I. Kanazawa. 1998. Distal sensory axonopathy after sarin intoxication. Neurology 51(4):1195-1197.

Howard, M.D., and C.N. Pope. 2002. In vitro effects of chlorpyrifos, parathion, methyl parathion and their oxons on cardiac muscarinic receptor binding in neonatal and adult rats. Toxicology 170(1-2):1-10.

IPCS (International Programme on Chemical Safety). 2001. Guidance Document for the Use of Data in Development of Chemical-Specific Adjustment Factors (CSAFs) for Interspecies Differences and Human Variability in Dose/Concentration Response Assessment. WHO/PCS/01.4. World Health Organization, Geneva [Online]. Available: http://www.who.int/pcs/harmon_site/ harmonize/docs/Guidance_Doc.pdf [accessed June 16, 2004].

Jarabek, A.M. 1995. Consideration of temporal toxicity challenges current default assumptions. Inhal. Toxicol. 7:927-946.

Kant, G.J., T.M. Shih, E.W. Bernton, H.G. Fein, R.C. Smallridge, and E.H. Mougey. 1991. Effects of soman on neuroendocrine and immune function. Neurotoxicol. Teratol. 13(2):223-228.

LaBorde, J.B., and H.K. Bates. 1986. Developmental Toxicity Study of Agent GB-DCSM Types I and II in NZW Rabbits, Final Report. AD-A168331/7. Prepared for U.S. Army Medical Research and Development Command, Fort Detrick, MD, by National Center for Toxicological Research, Jefferson, AR.

Langenberg, J.P., C. van Dijk, R.E. Sweeney, D.M. Maxwell, L.P. De Jong, and H.P. Benschop. 1997. Development of a physiologically based model for the toxicokinetics of C(+/−)P(+/−)−soman in the atropinized guinea pig. Arch. Toxicol. 71(5):320-331.

Luo, C., and J. Liang. 1997. Evaluation of the combined toxic effects of GB/GF and efficacy of Jielin injection against combined poisoning in mice. Toxicol. Lett. 92(3):195-200.

Manly, B.F.J. 2001. Statistics for Environmental Science and Management. Boca Raton, FL: Chapman & Hall/CRC.

Maxwell, D.M., C.P. Vlahacos, and D.E. Lenz. 1988. A pharmacodynamic model for soman in the rat. Toxicol. Lett. 43(1-3):175-188.

McCullagh, P., and J.A. Nelder. 1989. Generalized Linear Models, 2nd Ed. London: Chapman & Hall.

McDonough, J.H., and T.M. Shih. 1997. Neuropharmacological mechanisms of nerve agent-induced neuropathology. Neurosci. Biobehav. Rev. 21(5):559-579.

Metcalf, D.R., and J.H. Holmes. 1969. VII. Toxicology and physiology. EEG, psychological and neurological alterations in humans with organophosphorus exposure. Ann. N.Y. Acad. Sci. 160(1): 357-365.

Mioduszewski, R.J., S.A. Reuter, L.L. Miller, E.J. Olajos, and S.A. Thompson. 1998. Evaluation of Airborne Exposure Limits for G-Agents: Occupational and General Population Exposure Criteria. AD-A345618/3. ERDEC-TR-489. Edgewood Research, Development and Engineering Center, U.S. Army Chemical and Biological Defense Command, Aberdeen Proving Ground, MD [Online]. Available: http://chppm-www.apgea.army.mil/chemicalagent/caw/ERDEC-TR-489.pdf [accessed Sept. 22, 2004].

Mioduszewski, R.J., J. Manthei, R. Way, D. Burnett, B. Gaviola, W. Muse, S. Thomson, D. Sommerville, and R. Crosier. 2000. Estimating the probability of sarin vapor toxicity in rats as a function of exposure concentration and duration. Proceedings of the International Chemical Weapons Demilitarization Conference (CWD-2000), May 21-24, 2000, The Hague, Netherlands.

Mioduszewski, R.J., J. Manthei, R. Way, D. Burnett, B. Gaviola, W. Muse, J. Anthony, D. Durst, D. Sommerville, R. Crosier, S. Thomson, and C. Crouse. 2001. ECBC Low Level Operational Toxicology Program: Phase I-Inhalation toxicity of Sarin Vapor in Rats as a Function of Exposure Concentration and Duration. ECBC-TR-183. Edgewood Research Development and Engineering Center, Aberdeen Proving Ground, MD. August 2001.

Mioduszewski, R.J., J. Manthei, R. Way, D. Burnett, B. Gaviola, W. Muse, S. Thomson, D. Sommerville, and R. Crosier. 2002a. Interaction of exposure concentration and duration in determining acute toxic effects of sarin vapor in rats. Toxicol. Sci. 66(2):176-184.

Mioduszewski, R.J., J. Manthei, R. Way, D. Burnett, B. Gaviola, W. Muse, S. Thomson, D. Sommerville, R. Crosier, J. Scotto, D. McCaskey, C. Crous, and K. Matson. 2002b. Low-Level Sarin Vapor Exposure in Rats: Effect of Exposure Concentration and Duration on Pupil Size. ECBC-TR-235. Edgewood Chemical Biological Center, U.S. Army Soldier and Biological Chemical Command, Aberdeen Proving Ground, MD. May 2002.

Moore, D. 1998. Health effects of exposure to low doses of nerve agent- A review of present knowledge. Drug Chem. Toxicol. 21(Suppl. 1):123-130.

Morita, H., N. Yanagisawa, T. Nakajima, M. Shimizu, H. Hirabayashi, H. Okudera, M. Nohara, Y. Midorikawa, and S. Mimura. 1995. Sarin poisoning in Matsumoto, Japan. Lancet 346(8970):290-293.

Murata, K., S. Araki, K. Yokoyama, T. Okumura, S. Ishimatsu, N. Takasu, and R.F. White. 1997. A-symptomatic sequelae to acute sarin poisoning in the central and autonomic nervous system 6 months after the Tokyo subway attack. J. Neurol. 244(10):601-606.

Murphy, M.R., D.W. Blick, M.A. Dunn, J.W. Fanton, and S.L. Hartgraves. 1993. Diazepam as a treatment for nerve agent poisoning in primates. Aviat. Space Environ. Med. 64(2):110-115.

Myers, T.M., G. Galbicka, M.L. Sipos, S. Varadi, J.L. Oubre, and M.G. Clark. 2002. Effects of anticholinergics on serial-probe recognition accuracy of rhesus macaques (*Macaca mulatta*). Pharmacol. Biochem. Behav. 73(4):829-834.

Nakajima, T., S. Sato, H. Morita, and N. Yanagisawa. 1997. Sarin poisoning of a rescue team in the Matsumoto sarin incident in Japan. Occup. Environ. Med. 54(10):697-701.

Nakajima, T., S. Ohta, H. Morita, Y. Midorikawa, S. Mimura, and N. Yanagisawa. 1998. Epidemiological study of sarin poisoning in Matsumoto City, Japan. J. Epidemiol. 8(1):33-41.

Nakajima, T., S. Ohta, Y. Fukushima, and N. Yanagisawa. 1999. Sequelae of sarin toxicity at one and three years after exposure in Matsumoto, Japan. J. Epidemiol. 9(5):337-343.

Neter, J., M.H. Kutner, C.J. Nachtsheim, and W. Wasserman. 1996. Applied Linear Statistical Models, 4th Ed. Chicago, IL: Irwin Press.

Noort, D., A.G. Hulst, H.C. Trap, L.P. De Jong, and H.P. Benschop. 1997. Synthesis and mass spectrometric identification of the major amino acid adduct formed between sulphur mustard and haemoglobin in human blood. Arch. Toxicol. 71(3):171-178.

Noort, D., A.G. Hulst, and R. Jansen. 2002. Covalent binding of nitrogen mustards to the cysteine-34 residue in human serum albumin. Arch. Toxicol. 76(2):83-88.

NRC (National Research Council). 1982. Possible Long-Term Health Effects of Short-Term Exposure to Chemical Agents, Vol. 1. Anticholinesterases and Anticholinergics. Washington, DC: National Academy Press.

NRC (National Research Council). 1985. Possible Long-Term Health Effects of Short-Term Exposure to Chemical Agents, Vol. 3: Final Report, Current Health Status of Test Subjects. Washington, DC: National Academy Press.

NRC (National Research Council). 1989. Report of the Oversight Committee. Pp. 15-35 in Biologic Markers in Reproductive Toxicology. Washington, DC: National Academy Press.

NRC (National Research Council). 1993. Guidelines for Developing Community Emergency Exposure Levels for Hazardous Substances. Washington, DC: National Academy Press.

NRC (National Research Council). 1999. Review of the U.S. Army's Health Risk Assessment for Oral Exposure to Six Chemical Warfare Agents. Washington, DC: National Academy Press.

NRC (National Research Council). 2000. Strategies to Protect the Health of Deployed U.S. Forces: Analytic Framework for Assessing Risks. Washington, DC: National Academy Press.

NRC (National Research Council). 2001. Standing Operating Procedures for Developing Acute Exposure Guideline Levels for Hazardous Chemicals. Washington, DC: National Academy Press.

NRC (National Research Council). 2003. Acute Exposure Guideline Levels for Selected Airborne Chemicals, Vol. 3. Washington, DC: The National Academies Press.

NRC (National Research Council). 2004. Review of the Army's Technical Guides on Assessing and Managing Chemical Hazards to Deployed Personnel. Washington, DC: The National Academies Press.

O'Hern, M.R., T.R. Dashiell, and M.F. Tracy. 1997. Chemical defense equipment. Pp. 361-392 in Medical Aspects of Chemical and Biological Warfare, F.R. Sidell, E.T. Takafuji, and D.R. Franz, eds. Textbook of Military Medicine, Part 1. Warfare, Weaponry, and the Casualty, Vol. 3. Borden Institute, Walter Reed Army Medical Center, Washington, DC; Office of The Surgeon General U.S. Army, Falls Church, VA; U.S. Army Medical Department Center and School, Fort Sam Houston, TX; U.S. Army Medical Research and Materiel Command, Fort Detrick, Frederick, MD; and Uniformed Services University of the Health Sciences, Bethesda, MD.

Okudera, H. 2002. Clinical features on nerve gas terrorism in Matsumoto. J. Clin. Neurosci. 9(1):17-21.

Page, W.F. 2003. Long-term health effects of exposure to sarin and other anticholinesterase chemical warfare agents. Mil. Med. 168(3):239-245.

Piegorsch, W.W. 1998. Statistical aspects for combining information and meta-analysis in environmental toxicology. J. Environ. Sci. Health C Environ. Carcinogenesis Ecotoxicol. Rev. C16(2):83-89.

Polhuijs, M., J.P. Langenberg, and H.P. Benschop. 1997. New method for retrospective detection of exposure to organophosphorus anticholinesterases: Application to alleged sarin victims of Japanese terrorists. Toxicol. Appl. Pharmacol. 146(1):156-161.

Presidential Advisory Committee on Gulf War Veterans' Illnesses. 1996. Final Report. Washington, DC: U.S. Government Printing Office.

Ray, D.E. 1998. Chronic effects of low level exposure to anticholinesterases—a mechanistic review. Toxicol. Lett. 102/103:527-533.

Rengstorff, R.H. 1994. Vision and ocular changes following accidental exposure to organophosphates. J. Appl. Toxicol. 14(2):115-118.

Robbins, T.W. 2002. The 5-choice serial reaction time task: Behavioural pharmacology and functional neurochemistry. Psychopharmacology 163(3-4):362-380.

Robineau, P., Y. Leclercq, A. Gerbi, I. Berrebi-Bertrand, and L.G. Lelievre. 1991. An organophosphorus compound, VX, selectively inhibits the rat cardiac Na+, K(+)-ATPase alpha 1 isoform. Biochemical basis of the cardiotoxicity of VX. FEBS Lett. 281(1-2):145-148.

Romano, J., J. McDonough, R. Sheridan, and F. Sidell. 2001. Health effects of low-level exposure to nerve agents. Pp. 1-24 in Chemical Warfare Agents: Toxicity at Low Levels, S.M. Somani, and J.A. Romano, eds. Boca Raton, FL: CRC Press.

Senanayake, N., and L. Karalliedde. 1987. Neurotoxic effects of organophosphate insecticides: An intermediate syndrome. N. Engl. J. Med. 316(13):761-763.

Shih, M.L., J.D. McMonagle, T.W. Dolzine, and V.C. Gresham. 1994. Metabolite pharmacokinetics of soman, sarin and GF in rats and biological monitoring of exposure to toxic organophosphorus agents. J. Appl. Toxicol. 14(3):195-199.

Sidell, F.R. 1974. Soman and sarin: Clinical manifestations and treatment of accidental poisoning by organophosphates. Clin. Toxicol. 7(1):1-17.

Sidell, F. 1992. Clinical considerations in nerve agent intoxication. Pp. 156-194 in Chemical Warfare Agents, S.M. Somani, ed. San Diego, CA: Academic Press.

Sidell, F.R. 1997. Nerve agents. Pp. 129-180 in Medical Aspects of Chemical and Biological Warfare, F.R. Sidell, E.T. Takafuji, and D.R. Franz, eds. Textbook of Military Medicine, Part 1. Warfare, Weaponry, and the Casualty, Vol. 3. Borden Institute, Walter Reed Army Medical Center, Washington, DC; Office of The Surgeon General U.S. Army, Falls Church, VA; U.S. Army Medical Department Center and School, Fort Sam Houston, TX; U.S. Army Medical Research and Materiel Command, Fort Detrick, Frederick, MD; and Uniformed Services University of the Health Sciences, Bethesda, MD.

Sidell, F.R., and C.G. Hurst. 1997. The long-term health effects of nerve agents and mustard. Pp. 229-246 in Medical Aspects of Chemical and Biological Warfare, F.R. Sidell, E.T. Takafuji, and D.R. Franz, eds. Textbook of Military Medicine, Part 1. Warfare, Weaponry, and the Casualty, Vol. 3. Borden Institute, Walter Reed Army Medical Center, Washington, DC; Office of The Surgeon General U.S. Army, Falls Church, VA; U.S. Army Medical Department Center and School, Fort Sam Houston, TX; U.S. Army Medical Research and Materiel Command, Fort Detrick, Frederick, MD; and Uniformed Services University of the Health Sciences, Bethesda, MD.

Somani, S.M., and K. Husain. 2001. Low-level nerve agent toxicity under normal and stressful conditions. Pp. 83-120 in Chemical Warfare Agents: Toxicity at Low Levels, S.M. Somani, and J.A. Romano, eds. Boca Raton, FL: CRC Press.

Spencer, P.S., B.W. Wilson, and E.X. Albuquerque. 2000. Sarin other "nerve agents," and their antidotes. Pp. 1073-1093 in Experimental and Clinical Neurotoxicology, 2nd Ed., P.S. Spencer, and H.H. Schaumburg, eds. New York: Oxford University Press.

ten Berge, W.F., A. Zwart, and L.M. Appelman. 1986. Concentration-time mortality response relationship of irritant and systematically acting vapors and gases. J. Hazard. Mater. 13(3):301-309.

USACHPPM (U.S. Army Center for Health Promotion and Preventive Medicine). 2002a. Chemical Exposure Guidelines for Deployed Military Personnel. Technical Guide 230. U.S. Army Center for Health Promotion and Preventive Medicine. January 2002. [Online]. Available: http://chppm-www.apgea.army.mil/ deployment/ [accessed June 21, 2004]

USACHPPM (U.S. Army Center for Health Promotion and Preventive Medicine). 2002b. Chemical Exposure Guidelines for Deployed Military Personnel. A Companion Document to USACHPPM Technical Guide (TG) 230 Chemical Exposure Guidelines for Deployed Military Personnel. Reference Document (RD) 230. US. Army Center for Health Promotion and Preventive Medicine. January 2002.

U.S. Department of the Army. 1998. Risk Management. Army Field Manual 100-14. Headquarters, Department of the Army, Washington, DC. April 23, 1998.

van der Schans, G.P., D. Noort, R.H. Mars-Groenendijk, A. Fidder, L.F. Chau, L.P. de Jong, and H.P. Benschop. 2002. Immunochemical detection of sulfur mustard adducts with keratins in the stratum corneum of human skin. Chem. Res. Toxicol. 15(1):21-25.

van Helden, H.P.M., J.P. Langenberg, and H.P. Benschop. 2001. Low Level Exposure to GB Vapor in Air: Diagnosis/Dosimetry, Lowest Observable Effect Levels, and Performance Incapacitation. Final Report Grant Agreement DAMD17-97-1-7360. NTIS ADA400785. Prepared for U.S. Army Medical Research and Materiel Command, Fort Detrick, MD, by Prins Maurits Laboratory, TNO, Rijswijk, the Netherlands. April 2001.

van Helden, H.P.M., H.C. Trap, W.C. Kuijpers, B. Groen, J.P. Oostdijk, R.A.P. Vanwersch, I.H.C. Philippens, J.P. Langenberg, and H.P. Benschop. 2002. Low Level Exposure to GB Vapor in Air: Diagnosis/Dosimetry, Lowest Observable Effect Level, and Performance-Incapacitation. Operational Medical Issues in Chemical and Biological Defense [RTO-MP-075, AC/323(HFM-060)TP/37], North Atlantic Treaty Organisation, Research and Technology Organisation Symposium, 14-17 May 2001, Estoril, Portugal.

van Helden, H.P.M, H.C. Trap, J.P. Oostdijk, W.C. Kuijpers, J.P. Langenberg, and H.P. Benschop. 2003. Long-term, low-level exposure of guinea pigs and marmosets to sarin vapor in air: Lowest observable effect level. Toxicol. Appl. Pharmacol. 189(3):170-179.

Verbeke, G., and G. Molenberghs, eds. 1997. Linear Mixed Models in Practice: A SAS-Oriented Approach. Lecture Notes in Statistics 126. New York: Springer.

Wolfe, A.D., D.W. Blick, M.R. Murphy, S.A. Miller, M.K. Gentry, S.L. Hartgraves, and B.P. Doctor. 1992. Use of cholinesterases as pretreatment drugs for the protection of rhesus monkeys against soman toxicity. Toxicol. Appl. Pharmacol. 117(2):189-193.

Yanno, L.V., and Y.U. Musiichuk. 1997. Neuroparalytic agent acute poisoning and its long- term effects. [in Russian]. Med. Tr. Prom. Ecol. (6):5-7.

Yokoyama, K., S. Araki, K. Murata, M. Nishikitani, T. Okumura, S. Ishimatsu, N. Takasu, and R.F. White. 1998a. Chronic neurobehavioral effects of Tokyo subway sarin poisoning in relation to post-traumatic stress disorders. Arch. Environ. Health 53(4):249-256.

Yokoyama, K., S. Araki, K. Murata, M. Nishikitani, T. Okumura, S. Ishimatsu, and N. Takasu. 1998b. A preliminary study on delayed vestibulo-cerebellar effects of Tokyo Subway Sarin Poisoning in relation to gender difference: Frequency analysis of postural sway. J. Occup. Environ. Med. 40(1):17-21.

Young, R.A., D.M. Opresko, A.P. Watson, R.H. Ross, J. King, and H. Choudhury. 1999. Deriving toxicity values for organophosphate nerve agents: A position paper in support of the procedures and rationale for deriving oral RfDs for chemical warfare nerve agents. J. Hum. Ecol. Risk Asses. 5(3):589-634.

Yu, K.O, R.J. Mioduszewski, E.M. Jakubowski, S. Hulet, S.M. Thompson, and J.M. Gearhart. 2004. Preliminary development of physiologically-based pharmacokinetic/pharmacodynamic (PBPK/PD) model for low level exposure to chemical warfare agent (CWA) in mini pig. Abstract 2035. Toxicologist 78(S-1):419.

Abbreviations

α	alpha
AChE	acetycholinesterase
AEGL	acute exposure guideline level
β	beta
C	concentration
CNS	central nervous system
Ct	product of concentration and time
CV	conduction velocity
CWA	chemical warfare agent
DAAMS	depot area air monitoring system
DFP	diisopropyl fluorophosphates
DNA	dexyribonucleic acid
DOD	U.S. Department of Defense
EC_{01}	effective concentration estimated to be in 1% of the dosed animals
ECT_{50}	the concentration and time estimated to be that causes an effect in 50% of subjects
ED_{50}	dose estimated to be effective in 50% of the exposed individuals
EEG	electroencephalogram
FDA	Food and Drug Administration
GA	tabun
GB	sarin
GC	gas chromotography

GD	soman
GF	cyclosarin
GLP	good laboratory practice
HD	sulfur mustard
IU/L	international units per liter
L	liter
LC_{50}	concentration estimated to be lethal to 50% of subjects
LCt_{50}	lethal (estimated to be) concentration-time relationships in 50% of subjects
LD_{50}	dose estimated to be lethal to 50% of subjects
MAP2	mitogen-activated protein
MEGs	military exposure guidelines
mg	milligram
mL	milliliter
MOPP	mission-oriented protective posture
MRI	magnetic resonance imaging
mRNA	messenger ribonucleic acid
MS	mass spectroscopy
MTD	maximum tolerated dose
NRC	National Research Council
ORM	operational risk management
OP	organophosphate
PAC	presidential advisory committee
PBPD	physiologically based pharmacodynamic
PBPK	physiologically based pharmacokinetic
PON1	paraxonase 1
REM	rapid eye movement
RNA	ribonucleic acid
SFEMG	single fiber electromyography
t	exposure time
μg/kg	microgram per kilogram
μm	micrometer
μM	micromole
USACHPPM	United States Army Center for Health Promotion and Preventive Medicine
WPAFB	Wright-Patterson Air Force Base